T0344464

Climate Governance in China

This book explores how and why innovative climate policies spread across subnational regions and between governance levels in China.

Despite the significance of emerging economies in a pathway to a zero-carbon future, research to date on China's transformation governance remains limited. Drawing on a theoretical framework for policy diffusion and based on extensive data from expert interviews with Chinese decisionmakers and policy practitioners, Lina Li and Maia Haru Hall focus on the policy of emissions trading systems (ETS) and two key case studies: Shanghai and Hubei. The authors examine the role of the national government and how much freedom the subnational regions have in developing ETS policy, as well as pinpointing key actors and the role of policy and knowledge diffusion mechanisms. Overall, this book sheds light on the competition between China and the West in the transition to climate-friendly societies and economies, highlighting opportunities for cooperation between them.

This book will be of great interest to students and scholars of environmental politics and policy, climate change, urban studies, and Chinese studies more broadly.

Lina Li is the Asia Strategy Director at the International Energy Program (PIE), a platform hosted by the European Climate Foundation dedicated to securing a global clean power sector compatible with 1.5°C. With 15 years of experience in energy and climate policy, Lina has previously worked for various think tanks, consultancies, and NGOs, including Adelphi, Ecofys, and Greenovation Hub.

Maia Haru Hall is a consultant at Adelphi, a leading independent think tank for climate, environment, and development. She is also part of the Secretariat of the International Carbon Action Partnership, specializing in the Asia-Pacific region and working in carbon pricing and broader climate policy.

Routledge Focus on Environment and Sustainability

Participatory Design and Social Transformation
Images and Narratives of Crisis and Change
John A. Bruce

Collaborating for Climate Equity
Researcher—Practitioner Partnerships in the Americas
Edited by Vivek Shandas and Dana Hellman

Food Deserts and Food Insecurity in the UK
Exploring Social Inequality
Dianna Smith and Claire Thompson

Ecohydrology-Based Landscape Restoration
Theory and Practice
Mulugeta Dadi Belete

Regional Political Ecologies and Environmental Conflicts in India
Edited by Sarmistha Pattanaik and Amrita Sen

Circular Economy and the Law
Bringing Justice into the Frame
Feja Lesniewska and Katrien Steenmans

Land Tenure Reform in Sub-Saharan Africa
Interventions in Benin, Ethiopia, Rwanda, and Zimbabwe
Steven Lawry, Rebecca McLain, Margaret Rugadya, Gina Alvarado, and Tasha Heidenrich

For more information about this series, please visit: www.routledge.com/Routledge-Focus-on-Environment-and-Sustainability/book-series/RFES

Climate Governance in China

Policy Diffusion of Emissions Trading in Shanghai and Hubei

Lina Li and Maia Haru Hall

LONDON AND NEW YORK

from Routledge

First published 2024
by Routledge
4 Park Square, Milton Park, Abingdon, Oxon OX14 4RN

and by Routledge
605 Third Avenue, New York, NY 10158

Routledge is an imprint of the Taylor & Francis Group, an informa business

British Library Cataloguing-in-Publication Data
A catalogue record for this book is available from the British Library

ISBN: 978-1-032-35102-5 (hbk)
ISBN: 978-1-032-35103-2 (pbk)
ISBN: 978-1-003-32530-7 (ebk)

DOI: 10.4324/9781003325307

Typeset in Goudy
by Apex CoVantage, LLC

Contents

Acknowledgements

Lina Li

My utmost gratitude goes to Professor Dirk Messner (German Federal Environment Agency) for the mentorship, inspiration, and guidance. Also to Professor Haibin Zhang (Peking University) for his stellar work and review of this research.

The success of any empirical research on China depends on the cooperation of many people in the field. Special thanks to the nearly 60 people who generously gave their time and expertise to this research. Thank you to the three executive reviewers: Professor Maosheng Duan, Guoqiang Qian, and Dr. Xin Zhang. Many friends have also offered their insights and help along the way. Thank you to Axel Berger, Tobias Bernstein, Kang Chen, Matthias Stefan Falke, Kevin Osborne, Dongping Wang, Qinhong Wu, and Xuehua Zhang.

I thank my family for their enormous support. My parents and parents-in-law have shown faith in whatever I choose to do. My dear son, Linus, has profoundly changed my perspective on life and fueled my motivation to write this book. Most importantly, thank you to my husband Thomas Henze for his unending love and encouragement. It is to Linus and Thomas that I dedicate this work.

Maia Haru Hall

Thank you to my parents, Nao Nakanishi and Eric Hall, for instilling in me from the very beginning a true adoration of language and writing, alongside an ambition to act on climate change. I hope I grow up to be like you.

Thank you, of course, also to Lina. You are an incredible academic, colleague, woman, and friend.

Acronyms and abbreviations

BEE	Bureau of Ecology and Environment
CAFS	Chinese Academy of Fiscal Sciences
CBAM	Carbon border adjustment mechanism
CCD	Climate Change Division
CCEE	Climate Change and Energy Economics Study Center
CCER	Chinese Certified Emission Reductions
CCP	Chinese Communist Party
CDM	Clean Development Mechanism
CEC	China Electricity Council
CHEEX	China Hubei Carbon Emission Exchange
CHP	Combined heat and power
CMA	China Meteorological Administration
CO_2	Carbon dioxide
CQC	China Quality Certification Center
DAECC	Division of Atmospheric Environment and Climate Change
DEE	Department of Ecology and Environment
DRC	Development and Reform Commission
ECQT	Energy consumption quota trading
EDF	Environmental Defense Fund
ERI	Energy Research Institute
ETS	Emissions trading system
EU	European Union
FYP	Five-Year Plan
GDP	Gross domestic product
GHG	Greenhouse gas
GIZ	Gesellschaft für Internationale Zusammenarbeit
GW	Gigawatt
HXEE	Haixia Equity Exchange
ICAP	International Carbon Action Partnership
IEA	International Energy Agency
MEE	Ministry of Environment and Ecology

MIIT	Ministry of Industry and Information Technology
MoF	Ministry of Finance
MOST	Ministry of Science and Technology
MoT	Ministry of Transportation
MRV	Monitoring, reporting, and verification
NBS	National Bureau of Statistics
NCCC	National Coordination Committee on Climate Change
NCCP	National Climate Change Program
NCSC	National Center for Climate Change Strategy and International Cooperation
NDC	Nationally determined contribution
NDRC	National Development and Reform Commission
NEA	National Energy Administration
NGO	Non-governmental organization
NLGCC	National Leading Group on Climate Change
NPC	National People's Congress
OECD	Organization for Economic Cooperation and Development
OVUPRE	Wuhan Optics Valley United Property Rights Exchange
PMR	Partnership for Market Readiness
PRT	Pollution rights trading
RCEPD	Resource Conservation and Environmental Protection Division
RMB	Renminbi
SASAC	State-Owned Assets Supervision and Administration Commission
SECERC	Shanghai Energy Conservation and Emission Reduction Center
SECSC	Shanghai Energy Conservation Supervision Center
SEEE	Shanghai Environment and Energy Exchange
SICC	Shanghai Investment Consulting Company
SOE	State-owned enterprise
SPF	Strategic Programme Fund
tce	Ton of coal equivalent
tCO2e	Ton of CO_2e
UK	United Kingdom
UN	United Nations
USA	United States of America
WBGU	Wissenschaftlicher Beirat der Bundesregierung Globale Umweltveränderungen (German Advisory Council on Global Change)
ZEIC	Zhejiang Economic Information Center

1 Introduction

The devastating impacts of climate change are threatening our continued existence. Our current development pathway must transform entirely away from fossil fuels—and fast. But what exactly is transformation? Since the end of the Cold War, the international system has changed fundamentally (Godehardt, 2016; Messner, 2015; WBGU, 2011). Not only is the global economy increasingly interconnected, but diffuse architectures of power also proliferate. Political, economic, and environmental systems are increasingly complex. We are now in the Anthropocene Era, where humans have become what drives the evolution of our planetary ecosystem.

Transformation is understood to exist under the framework of such an international system. With climate change and other environmental degradation poised as the greatest challenges of our time, we must urgently develop a social contract that facilitates transformation away from the high-carbon and quantity-driven economic models that take center stage today.

The alternative must be sustainable, climate friendly, and quality driven. The level of transformation necessary to achieve this is extremely challenging, requiring concerted action and without the luxury of time. The change must bring together all countries, from the most industrialized economies, to those emerging, to those with the least resources (WBGU, 2011). From a governance perspective, this transformation is a process embedded in the global multilevel political architecture, concerning myriad public, private, and civil society actors with different (and sometimes competing) interests, beliefs, and resources (ibid.). The success of such a transformation thus requires unwavering commitment and calls for brave new leaders.

A global shift in power is already underway. The economic and political rise of emerging economies has been a salient feature of global governance over the past two decades (Berger, 2015). Western countries have weathered various storms: financial, migration, and public health crises, against a backdrop of swelling support for populist parties. On the other hand, emerging economies have been leading global recovery and growth, working to enhance their individual and collective power. The rise of countries like Brazil, India,

DOI: 10.4324/9781003325307-1

South Africa, and particularly China has challenged the post–Cold War status quo not just economically, but increasingly also politically.

China's role in the global transformation

Among them, home to almost one quarter of the world's population and the fastest-growing economy on the planet, is China. As the world's largest emitter of carbon dioxide (CO_2), it also holds great significance in shaping this global transformation. But China is at a crossroads. Its leadership under Xi Jinping has seen the introduction of several new concepts, including ecological civilization, which highlights respect for nature and the environment. These signal the country's grand ambitions as it looks to the future. But it is the at-home struggles closely felt by the public, such as air quality, food, water, and health, that continue to put pressure on government leaders. Public concerns over environmental problems have been growing since the 2000s, alongside the evolution of the crisis itself. There are ever more venues for people to raise and engage with these issues, including a booming social media landscape. Though this can push the government towards committing to new environmental protection targets and policies, the power of growing environmental movements has also caused Chinese leaders concern over social and political unrest; this has led to renewed control of the online and physical space for advocacy.

China has indeed shown willingness to combat environmental crises and climate change. The government is carefully studying and testing new concepts and instruments, working on many fronts to mitigate emissions and transform the country's energy system (State Council, 2008; Greenovation Hub, 2012). It has made significant progress in the past decade, its green transition intertwined with much political and economic reform (Greenovation Hub, 2014). Internationally, China's increasing weight in the global power structure is generating new demand for it to take on more responsibility. Accordingly, Chinese policy elites are redefining the country's role in the global arena and how it interacts with others. Climate change is thus a topic that has both internal and external relevance for China. This book examines its domestic dimension.

Lack of transformation governance research on China

The scale and speed of transformation around the world that has already begun does not necessarily point to its success (WBGU, 2011). Research to date has concentrated primarily on Western industrialized societies. However, emerging countries will be decisive in the success or failure of the global transformation journey (ibid.). This book focuses on China as perhaps the world's most significant emerging country, more specifically its subnational transformation governance. It provides insights into the Chinese governance structure and policymaking process.

Traditionally, Western literature has assumed the federal system associated with electoral democracy as the starting point for research. Though further assessment concludes that in federal democratic countries, policy diffusion is more likely to be horizontal and learning-oriented (Graham et al., 2013), policy diffusion tends to be more vertical and coercion/order-oriented in authoritarian or unitary state countries (Heilmann, 2008). Some scholars even argue that policy diffusion is premised on decentralization and that centralization is the enemy of diffusion (Shipan & Volden, 2012). This certainly does not reflect the reality of China, an authoritarian country with many examples of policy diffusion and dynamic policy innovation.

A generous body of literature on policy diffusion and innovation since the 2000s focuses on the motivation, content, and characteristics of subnational government policy innovation (K. P. Yu, 2010; He, 2011; X. D. Yang, 2011; Wu et al., 2007; Chen & Wang, 2014); others on policy innovation sustainability (X. H. Yu, 2013; Liu, 2014); and yet others use quantitative methods to carry out empirical analyses and test the applicability of classical policy diffusion theory on China. Though Zhou (2012) posits that most classical policy diffusion models in China begin with experiment and end in rollout, they do not unpack the specific mechanisms at play. Policy innovation and diffusion at the subnational level in China is yet to be fully understood (Zhang, 2015, p. 79).

There have also been several recent attempts made by Chinese researchers to develop their own analytical frameworks to study policy diffusion in China. We refer especially to the diffusion frameworks developed by Yang Hongshan (2013) and Yang Daifu (2016). The former proposed the "dual-track policy experimentation" model, where Chinese national decisionmakers give selected regions the right to experiment with and implement policy, creating a structure with parallel policy tracks (pilot alongside non-pilot regions). The latter built on this and developed an analytical diffusion framework covering actors from several levels, national or regional exchange, and oblique influence. Daifu also explored influencing mechanisms including coercion, inducing, learning, competition, emulation, and socialization. However, neither applied their framework to empirical studies. This book develops its own framework based on these earlier versions from both Western and Chinese scholars, testing it on empirical case studies in China.

Does China's governance model facilitate modernization?

The debate on China's governance model and whether it is fit for the country's modernization—and transformation—to a carbon-neutral future is ongoing. China's central government–oriented structure is a distinctive governance approach. Some have analyzed the country's climate governance through the authoritarian environmentalist lens (Gilley, 2012; Shearman and Smith, 2007). Others identify its climate governance as "top-down" (Qi, 2011) or state-signaling (Ran, 2013), steering policy implementation from above by providing targets and guidelines for local governments. However, this top-down label is only part of the picture. Recent years have seen the emergence of bottom-up initiatives from China's provinces and cities (E3G, 2008; Qi et al., 2008; Schröder, 2012). Against this backdrop, this book contributes to the understanding of the dynamic interactions across and within different political levels.

The international literature on both governance theory and environmental and energy politics is historically dominated by advocacy for "decentralization". Decentralization is portrayed as the new solution to economic and administrative public demand, a normative institutional design fit for the modern state (G. C. Chen, 2016). However, there is an emerging recognition of the effectiveness of the more centralized approach, that the prescription should not be uniform to all polities (Hutchcroft, 2001) but should focus on specific national and policy contexts. This book reveals the opportunities provided by China's governance structure in promoting policy cooperation, mutual learning, and the deployment of innovative policies.

China wields decisive influence on whether the world can successfully embrace a carbon-neutral future. Yet existing literature on the country's transformation focuses on specific policies and measures and their effectiveness, comparing them with those in other countries. Very few explore the transformation policymaking process and how it evolves from a governance perspective. Further investigation into governance models beyond liberal democracy is necessary to grasp transformation on a global scale.

> Whether democratic regimes are better suited to foster sustainable development than autocratic regimes is an ongoing debate. Empirical studies yield mixed results, which suggest that democracies do not outperform autocracies with regards to development results, for instance . . . climate-friendly policies.
>
> (TWI2050—The World in 2050, 2018, p. 108)

We must move beyond traditional dichotomies and understand which overlaps between different governance structures can spur a faster and more sustainable transformation.

The fundamentals of policy diffusion

Mainstream policy diffusion theory has a long tradition in social and public policy sciences, especially the comparative politics analysis of the US federal system (Graham et al., 2013). The classical definition of policy diffusion is "when a government implements a new policy or project for the first time—regardless of if it has been implemented in other times or places" (Walker, 1969, p. 881).

Such a general definition focusing on the outcome of policy adoption has gradually become more targeted, focusing on "interdependency" (see Collier & Messick, 1975; Graham et al., 2013; Heinze, 2011; Simmons et al., 2006). Here, policy diffusion occurs when one government's decision about whether to adopt a policy innovation is influenced by previous choices by other governments in an *interactive* process (Braun & Gilardi, 2006). The decision-making party makes an "interdependent, but uncoordinated, decision" (Elkins & Simmons, 2005, p. 35) regarding a policy that others have pursued previously. The study of policy diffusion reveals what is required to ensure that innovative policies are adopted (Jordan & Huitema, 2014, p. 729).

Over the past half century, policy diffusion theory has evolved by drawing on theoretical schools in political science, public policy, social science, and international relations. "Theories of diffusion encompass a wide array of assumptions about who the primary actors are, what motivates their behavior . . . and their ultimate goals" (Simmons et al., 2006, p. 787). These are presented as diffusion mechanisms, which are "a systematic set of statements that provide a plausible account of how [two variables] are linked" (Hedstrom & Swedberg, 1998, p. 7). That is, they can explain why and how the policy of government A influences that of government B.

Mainstream policy diffusion literature today identifies four main diffusion mechanisms: coercion, competition, learning, and emulation (Gilardi, 2016; Simmons et al., 2008b; Shipan & Volden, 2008). Heinze (2011) adds one more: socialization, or how interaction leads to the development and internalization of normative ideas (ibid.). Socialization has also been highlighted by some Chinese scholars (e.g., Yang, 2016) in their diffusion frameworks. This book thus explores these five major mechanisms: coercion, competition, learning, emulation, and socialization, in addition to a final mechanism—indirect diffusion—that emerges from the empirical data.

Mechanisms of policy diffusion

The **coercive policy diffusion mechanism** describes when powerful jurisdictions force others to adopt a policy via conditionality or other forms of enforcement. Through coercion, policies in an asymmetric power structure spread vertically from the center to the periphery (Simmons et al., 2008a, p. 10). This vertical relationship distinguishes coercion from the other diffusion mechanisms that tend to be more decentralized (Berger, 2015, p. 62). Horizontal coercion may also occur across polities such as states or localities (Shipan & Volden, 2008).

The concept of coercion can be difficult to apply empirically. Firstly, "it is necessary to identify the coercive actors, to show that they promote the policy in question, and to show evidence that their promotion increases the likelihood of policy adoption" (Dobbin et al., 2007, p. 457). Furthermore, the weaker actor must enforce policies, regardless of any opposition (Berger, 2015, pp. 63–64). The more powerful actor can inject influence into the less powerful actor via positive or negative incentives—or carrots and sticks (Heinze, 2011, p. 17). Coercion may also be exercised through physical force, the manipulation of economic costs and benefits, or the monopolization of information or expertise (Dobbin et al., 2007). D. F. Yang (2016) identifies a non-exhaustive list of ways coercion exists in the Chinese governance structure context:

1 Specific, concrete commands and target-setting: an order is given to the subordinate government which could take the form of guidelines, policies, or decisions made by the central or higher party committee of the Chinese Communist Party (CCP); or executive regulations, rules, decisions, or "red-letter head documents" from the central or the higher-level government;
2 More macrolevel and longer-term directional guidance from the central or higher-level government, including speeches or instructions from leadership;
3 Recognition of certain policy experience or innovation by high-level government or party officials; or
4 Financial and other incentives or support.

Like coercion, the **competition mechanism** is based on setting positive and/or negative incentives for policy adoption, manipulating and influencing a jurisdiction's utility calculations (Heinze, 2011, p. 17). This type of interdependence mirrors the prisoner's dilemma: "cooperation might lead to regulatory policies that make all better off, but there is a constant temptation to adopt regulatory policies that improve one's own standing" (Lazer, 2001, p. 476). Policy choices thus create externalities for those in the same competition space (Braun & Gilardi, 2006). These put adaptive pressure on jurisdictions by altering the material payoff structure associated with pursuing a specific policy. Policymakers hence monitor the policies of jurisdictions they consider to be competitors, anticipating or reacting to changes that may deteriorate the competitive position of their own jurisdiction (Gilardi & Wasserfallen, 2019).

Competition diffusion in the Chinese political system is distinct from the competition for votes in Western democratic political systems and the dominance of economic competition in the classical diffusion literature, where countries compete for capital and export markets. In China, structural competition exists among subnational governments for political resources and recognition from the central government (Wang & Lai, 2013).

The **learning mechanism** is when the experiences of external actors, or the success (or failure) of certain policies are used to provide effective solutions to the policy follower region's own domestic issues. Some scholars argue that policymakers may also be interested to learn about political viability, implications for reelection, reappointment, or promotion (Graham et al., 2013). Others point to the different elements of a policy's success, including policy goals, implementation challenges, and political support (Maggetti & Gilardi, 2016).

There are different forms of learning. Rational learning assumes that policymakers systematically assess information on policy consequences without bias or discrimination (Meseguer, 2006). A more realistic characterization of rational learning, however, builds on the Bayesian updating analogy, that policymakers in one jurisdiction have certain priors on a policy's likely effects and then update them based on the information from the experience of other jurisdictions. A more recent strand of the literature deviates further from the assumption of rational learning by studying "bounded" learning processes in which policymakers rely on cognitive shortcuts when gathering and assessing policy information (Weyland, 2005; Bamert et al., 2015; Berger, 2015). There is, however, no absolute distinction between rational and bounded learning, especially at the empirical level. No actor is expected to have full rationality in practice, and it is difficult to prove that policymakers have truly systematically analyzed a policy's effectiveness.

The **emulation (imitation) mechanism** conceptualizes diffusion as a process whereby policies spread because they become socially accepted rather than as a result of cost-benefit calculations. The legitimacy of policy choices increases via the imitation of policy solutions found elsewhere (Bennett, 1991; DiMaggio & Powel, 1991; Elkins & Simmons, 2005). Typically, the adoption of a policy by a leading jurisdiction provides the critical impetus for emulation diffusion. In contrast to learning, policymakers' motivation to adopt a policy is not its success but that it has been carried out by other jurisdictions perceived as leaders. Where learning focuses on the action, emulation focuses on the actor (Shipan & Volden, 2008).

The constructivist school also often refers to the epistemic communities of policy experts as an important impetus for emulation or learning diffusion via a rationalization process (Berger, 2015, p. 78). Similarly, policy experts can trigger softer forms of coercion such as hegemonic ideas.[1] These mechanisms differ based on whether the experts are independent (emulation or learning) or if they are assigned by a more powerful actor (softer forms of coercion) (Simmons et al., 2008a, p. 26; Berger, 2015, p. 78). Another difference is the attitude of the policy follower region—whether it is willing to adopt the policy (emulation or learning) or opposes it (coercion).

Empirical evidence points to the limited impact of emulation in terms of the depth of change (Heinze, 2011, p. 22). Emulation may be more influential in agenda-setting rather than domestic decision-making (Cohen-Vogel & Ingle, 2007).

The **socialization mechanism** endorses the same assumption as emulation that constructed ideas and norms about a policy's appropriateness foster its diffusion (Heinze, 2011, p. 19). Socialization suggests that interaction between actors leads to the development and internalization of normative ideas. This then further shapes their perceptions on the legitimacy of certain norms and policies and results in a redefinition of their identities and belief systems (ibid.). Here, "entrepreneurs" again serve as powerful agents in the

Table 1.1 Macrolevel characteristics of policy diffusion mechanisms

	Coercion	*Competition*	*Learning*	*Emulation*	*Socialization*
School of thought	Rationalism; realism	Rationalism; liberal institutionalism	Rationalism; liberal institutionalism	Constructivism	Constructivism
Trigger	Material consequence	Material consequence	Material consequence	Ideational impact	Ideational impact
Pace of adoption	Medium/Low	High	Medium/Low	High	Medium
Variation of content	Low	Low	High/Medium	Low	Medium
Length of policy impact	Medium/Short	Long	Long	Short	Long
Sensitivity to domestic policymaking	Uncertain	Low	High	Low	Low

Source: Author, drawing on Heinze (2011); Berger (2015)

socialization process (Finnemore & Sikkink, 1998; Mintrom, 1997; Teodoro, 2009). Efficacy distinguishes socialization from learning. If a policy's efficacy has influenced its adoption by another jurisdiction, then learning has occurred (Simmons et al., 2008a, p. 35).

How the mechanisms differ

Despite broad consensus about the main policy diffusion mechanisms, concrete operationalization and differentiation are still disputed (Gilardi, 2016). Diffusion mechanisms can be distinguished by criteria that draw on patterns at the macrolevel.

Important macrolevel characteristics include the type of trigger, the pace of adoption of new policies, the adaptation of policy design to the recipient's local conditions, whether longer-term policy impacts are exhibited, and "sensitivity" to domestic policymaking (see Table 1.1).

Coercion, competition, and learning follow rationalist theories, while emulation and socialization follow the constructivist tradition. According to rationalism, actors are supposed to maximize utility; for constructivists, norms and beliefs about appropriateness shape actors' behavior (Heinze, 2011). The trigger for policy diffusion hence can be clustered into two groups: one relates to material preferences, interests, desires, and consequences and the other to social expectations, norms, values, and rule-driven behavior. In other words, one group is functional value-driven and the other ideational value-driven (ibid.).

How quickly the recipient government takes up the policy from elsewhere is also critical. This "speed of diffusion" is defined not in absolute terms but by comparing it across different periods or policy follower regions. Competition and emulation see a high speed of diffusion, socialization a medium pace, and learning and coercion a slower pace. The duration of the policy impact once injected by the policy supplier towards the recipient or follower government is relative. Competition, learning, and socialization are expected to see longer lasting impact than coercion and emulation.

The extent to which a policy's design and implementation are adjusted to the recipient region's local conditions also helps distinguish between the mechanisms. Low variation of policy content is expected for coercion, competition, and emulation, medium variation for socialization, and high to medium for learning.

Finally, sensitivity to the domestic policymaking process also differs, i.e., interactions between external forces and domestic factors of the policy recipient. According to the conceptual logic of the competition, emulation, and socialization mechanisms, domestic factors play a limited role, but a more prominent one in learning. More specifically, for competition, the recipient government adopts a certain policy to "win" the competition against others. In the case of the Chinese regions, this could be political performance such as the realization of energy-saving or emissions-reduction targets set by the Five-Year Plans (or FYPs, China's blueprints for economic and social development) (Qi, 2011). Another aim for competition in China is to gain political credibility in the eyes of the central government. In the emulation mechanism, the recipient government takes the policy supplier region as a role model and mimics their policy. In the socialization mechanism, the recipient government adopts the policy to show that they adhere to certain advanced norms such as low-carbon development. The learning mechanism, in contrast, is based on a cost-benefit analysis of the policy, which usually involves domestic stakeholder engagement (Berger, 2015, p. 72).

There is no clear-cut distinction between the mechanisms. Actors may also incorporate social and ideational values into their utility calculations, complicating things further (Heinze, 2011, p11). We must therefore operationalize the diffuse mechanisms to form the analytical framework that guides the collection of empirical data and this book's analysis.

Operationalizing the mechanisms

The definitions of the main diffusion mechanisms are broad and thus require operationalization. As Maggetti and Gilardi (2016) note, "the priority should be to construct a good measure for the main mechanism under consideration, rather than including as many mechanisms as possible" (p. 103). Accordingly, we use one major operationalization model per mechanism. For simplification, the graphic illustrations of the operationalized diffusion mechanisms focus on policy adoption as the outcome of the diffusion process. This operationalization draws on previous work by Beach and Pedersen (2013) and Berger (2015) and develops five models for the specific purpose of analyzing transformative climate policy diffusion in China.

Coercion can occur both vertically and horizontally, based on an asymmetric distribution of power. The process begins with the initiative of the stronger actor, i.e., the policy supplier region or jurisdiction, who provides either positive or negative incentives to promote a certain policy to the weaker actor, i.e., the policy follower region (see Figure 1.1). Domestic groups in the policy follower region may oppose the objectives (ends) or measures (means) of the policy. This opposition may be moderate or even "invisible" for those outside of the policymaking process; this is particularly true in China where the central government yields significant power and where decision-making often takes place behind closed doors. In this context, stakeholder concerns may be expressed to policymakers only via closed channels. Incentives can manipulate the cost-benefit structure and lead to the acceptance and adoption of the policy or some of its elements in the policy follower region.

The **competition** process is also triggered by a change in utility calculations resulting from positive/negative incentives. However, it typically begins with the policy follower (rather than supplier) region (see Figure 1.2). Policymakers in the

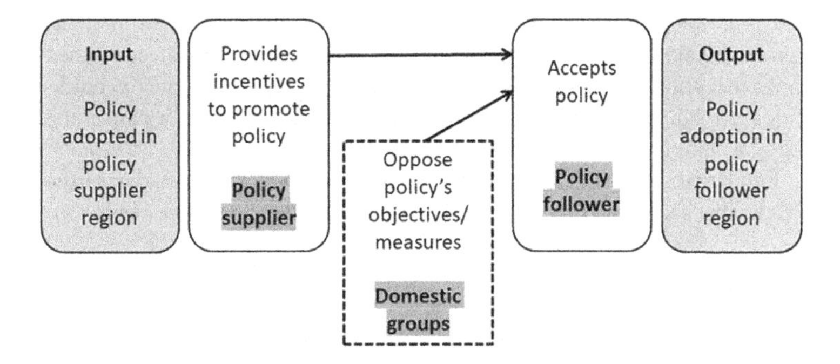

Figure 1.1 Operationalizing the coercion mechanism.

Source: Author

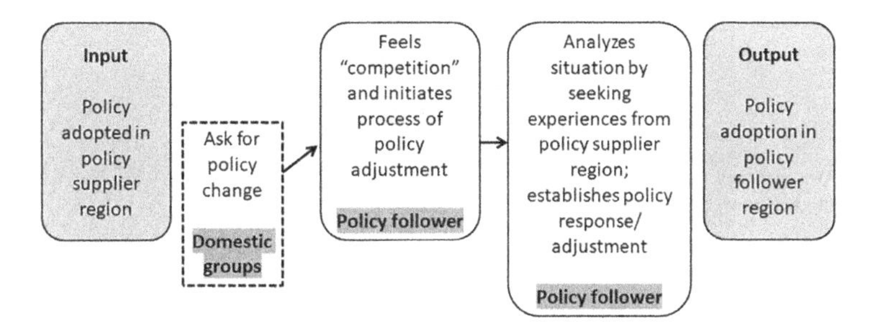

Figure 1.2 Operationalizing the competition mechanism.

Source: Author

follower region monitor the policy choices of jurisdictions they consider to be competitors. These create externalities that cause pressure to adjust or react proactively. In this initial phase, there may be domestic groups advocating for the necessity of policy adjustment. Policymakers in this region invest resources to analyze the situation. This can result in the adoption of a similar policy, as the goal becomes to not lose out on—or to even win—the perceived competition. This concept of "not losing" is very important in the Chinese context, where often it is not about being the winner but rather staying within a "safe" range regarding policy performance.

The **learning** process may start in either policy supplier or follower region (see Figure 1.3). A follower region may observe the success of a policy adopted in the supplier region, which then may kickstart its own learning process via information gathering and experience exchange. Alternatively, the supplier region may actively promote its success, triggering a similar process.

Learning involves the study of policy experience and cost-benefit analysis. Domestic stakeholders often articulate their preferences and concerns based on local context. The process can be complex, so careful government coordination is required. The policy follower region adopts the whole or part of the policy, adapting design or implementation to local conditions.

Emulation begins with a supplier region (or a group of regions) considered to be a leader that adopts a policy, which it may also choose to promote (see Figure 1.4).

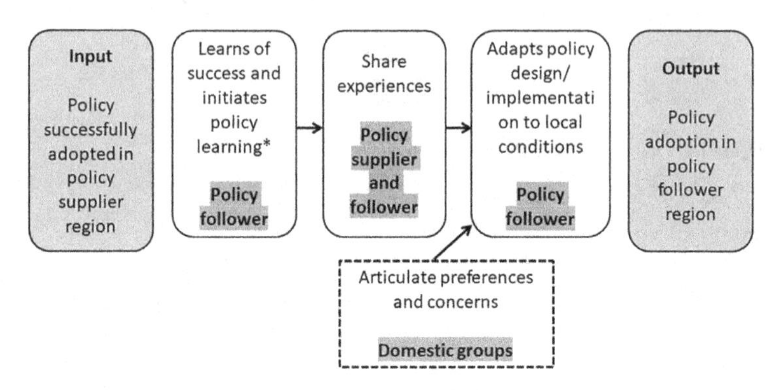

Figure 1.3 Operationalizing the learning mechanism.

Source: Author

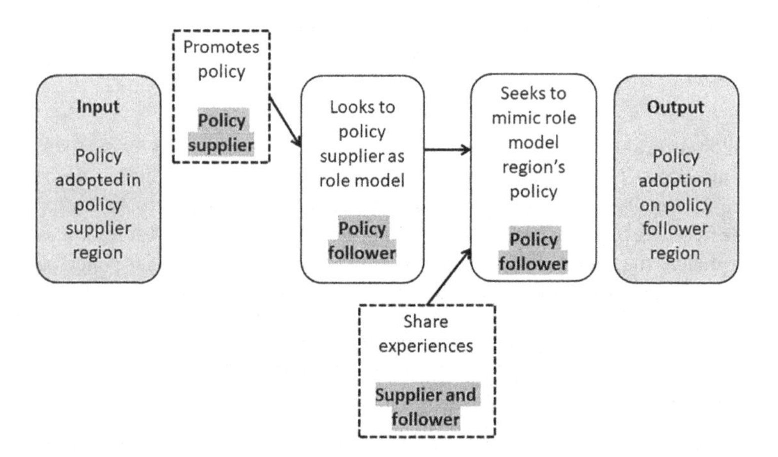

Figure 1.4 Operationalizing the emulation mechanism.

Source: Author

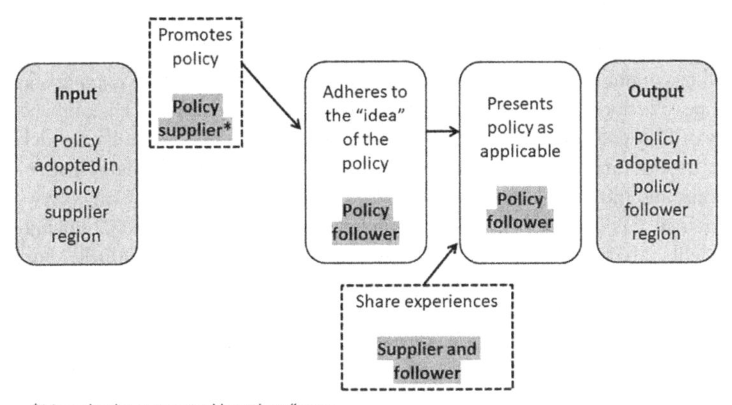

Figure 1.5 Operationalizing the socialization mechanism.

Source: Author

The follower region looks to the leader and subsequently mimics said policy. They may also share their experiences, a process often initiated by the follower region. The follower region copies the policy in entirety or in part from the leading region, without substantial adjustment.

Like emulation, the **socialization** mechanism is based on a normative trigger. The policy supplier region or other "norm entrepreneurs" such as national or regional government or policy experts promote a certain policy (see Figure 1.5). The policy follower region then adheres to this policy idea, based on belief in its ideational value. The supplier and follower regions may also exchange experiences. The follower region presents the policy to external or internal audiences (or both) as appropriate, resulting in policy adoption in entirety or in part.

About this book's approach

"How and why do low-carbon transformation policies spread across different regions and between governance levels in China?" This question is the starting point for this book, which seeks to explore the diffusion mechanisms that best explain how Chinese regions introduce climate policy, the key players in the diffusion process and how they participate, and the main characteristics of Chinese climate policy diffusion. The analytical policy diffusion framework is well suited to investigating China's transformation governance. It allows us to test several possible explanations for the dynamic development of the recent transformation.

This book employs mixed methods, combining a literature review,[2] semi-structured interviews, and surveys. For any political process, tracing it is critical to conduct a broad base of expert interviews to allow for the comparison of perspectives, though a degree of subjectivity is unavoidable. The complexity of ETS policymaking that involves a spectrum of actors also calls for such a range of interviewees. However, obtaining access to relevant data, institutions, and individuals is a huge obstacle to empirical studies on China's governance. The authors' existing ETS and climate policy network in China and a deliberate snowballing method have been used overcome this hurdle. Furthermore, the opaqueness of policymaking in China translates into a widespread reluctance of policymakers to engage with foreign or foreign-based researchers. The authors' *Guanxi* (the Chinese term for relationships) developed through years of professional working experience in the field of climate change—and ETS—policy allowed for this to be circumvented in part. Still, difficult-to-reach government representatives account for only around 10% of total interviews.

In total, 58 individuals were interviewed.[3] All seven pilot regions plus 11 non-pilot regions (accounting for half of the regions in China) and all five key national technical supporting institutions are represented.

Two case studies form the core of this book's analysis: the diffusion of emissions trading in Shanghai and Hubei. The empirical data used focuses on the piloting of such emissions trading systems (ETS) (or "carbon markets"). Process tracing is used to analyze diffusion mechanisms that led to the introduction of climate policies across regions and between governance levels over time. ETS policy in China is still evolving. This book draws on developments that occurred between 2011 and 2019, in the lead-up to the national ETS initially expected to begin operating by 2020.[4]

The remainder of this book is organized as follows. Chapter 2 outlines the general context of the empirical study: ETS in China. It introduces the Party State and the Tiao-Kuai system that underpin China's governance structure. Then, it discusses the broader debate of China's transformation governance, providing an overview of the country's climate policy developments and of the seven regional ETSs, before introducing the legal and institutional framework underlying China's climate and ETS policy. It delves then into the development of the national ETS and the selection of the two case study regions. The following two chapters unpack the policy diffusion experiences of Shanghai and Hubei. Each contains a brief background of the region's economic and social context, a detailed overview of the pilot's governance structure, its evolution over time, and a detailed analysis of the various diffusion mechanisms. Chapter 5 offers a conclusion on the role of policy diffusion in China's transformation via a comparison of the case studies, how this differs from Western democracies, and the enabling factors for successful climate policy diffusion. It also discusses the conceptual and practical implications for China's transformation. Finally, Chapter 6 delves into continued top-down influences, future prospects for the regional ETSs, and developments triggered by other carbon pricing progress around the world.

Notes

1 The thrust of hegemonic ideas is that dominant ideas become rationalized, often with elegant theoretical justifications, and influence how policymakers conceptualize their problems and order potential solutions (Dobbin et al., 2007).
2 A review of Chinese and English academic literature, government documents, policy reports, statements from government officials, media reports, and other grey literature.
3 Of this, six with policymakers, 41 with policy experts from former government, academia, consultancies, exchanges, SOEs, or technical supporting institutions, eight with industry stakeholders, two with non-governmental organizations (NGOs), and one with a media representative. A coding system is used to reference interviews, comprising three parts: (1) region, i.e., region's initials or *N* for interviewees from (inter)national institutions; (2) institution, i.e., *G* = government, *E* = expert, *P* = private sector, *N* = NGO, *M* = media; and (3) date of interview six digits, i.e., ddmmyy. For example, SH-E-271119 is an interview conducted with an expert from Shanghai on 27 November 2019.
4 This was postponed to 2021 largely due to the COVID-19 pandemic.

References

Bamert, J., Gilardi, F., & Wasserfallen, F. (2015). Learning and the diffusion of regime contention in the Arab Spring. *Research & Politics*, 2(3), 1–9.
Beach, D., & Pedersen, R. B. (2013). *Process-Tracing Methods: Foundations and Guidelines*. Ann Arbor, MI: University of Michigan Press.
Bennett, C. J. (1991). Review article: What is policy convergence and what causes it. *British Journal of Political Science*, 21(2), 215–233.
Berger, A. (2015). *Swimming with the Tide: China and the Protection of Foreign Direct Investment*. DPhil thesis, Universität Duisburg-Essen: Duisburg.
Braun, D., & Gilardi, F. (2006). Taking 'Galton's problem' seriously: Towards a theory of policy diffusion. *Journal of Theoretical Politics*, 18(3), 298–322.
Chen, G. C. (2016). *Governing Sustainable Energies in China*. Cham, Switzerland: Springer International Publishing AG.
Chen, J. X., & Wang, Y. C. 陈家喜 汪永成. (2014). 政绩驱动: 地方政府创新的动力分析 [Political achievement driven: An analysis of the dynamics of local government innovation]. *Zhengzhixue yanjiu*, 4, 50–56.
Cohen-Vogel, L., & Ingle, K. (2007). When neighbours matter most: Innovation, diffusion and state policy adoption in tertiary education. *Journal of Education Policy*, 22(3), 241–262.
Collier, D., & Messick, R. E. (1975). Prerequisites versus diffusion: Testing alternative explanations of social security adoption. *American Political Science Review*, 69(4), 1299–1315.
DiMaggio, P. J., & Powell, W. (1991). The iron cage revisited: Institutionalized isomorphism and collective rationality in organizational fields. In P. J. DiMaggio & W. Powell (Eds.), *The New Institutionalism in Organizational Analysis*. Chicago: University of Chicago Press, 63–82.

Dobbin, F., Garrett, G., & Simmons, B. (2007). The global diffusion of public policies: Social construction, coercion, competition, or learning? *Annual Review of Sociology*, 33, 449–472.

E3G. (2008, November). Low carbon. zones: Case studies of potential areas for closer EU-China cooperation. *Matt Findlay, Taylor Dimsdale & Shin Wei Ng*. www.e3g.org/library_asset/low-carbon-zones-case-studies-of-potential-areas-for-closer-eu-china-cooper/

Elkins, Z., & Simmons, B. (2005). On waves, clusters and diffusion: A conceptual framework. *Annals of the American Academy of Political and Social Science*, 598, 33–51.

Finnemore, M., & Sikkink, K. (1998). International norm dynamics and political change. *International Organization*, 52(4), 887–917.

German Advisory Council on Global Change (WBGU). (2011). Welt im Wandel: Gesellschaftsvertrag fuer eine Grosse Transformation [World in transition: A social contract for sustainability]. *WBGU Flagship Report*. www.wbgu.de/de/publikationen/publikation/welt-im-wandel-gesellschaftsvertrag-fuer-eine-grosse-transformation

Gilardi, F. (2016). Four ways we can improve policy diffusion research. *State Politics & Policy Quarterly*, 16(1), 8–21.

Gilardi, F., & Wasserfallen, F. (2019): The politics of policy diffusion. *European Journal of Political Research*, 58(4), 1245–1256.

Gilley, B. (2012). Authoritarian environmentalism and China's response to climate change. *Environmental Politics*, 21(2), 287–307.

Godehardt, N. (2016). No end of history: A Chinese alternative concept of international order? *SWP. Research Paper 2016/RP 02*, Berlin.

Graham, E. R., Shipan, C. R., & Volden, C. (2013). The diffusion of policy diffusion research in political science. *British Journal of Political Science*, 43(3), 673–701.

Greenovation Hub. (2012, April). The first year of China's twelfth five year plan: Success or failure for climate change efforts? *Annual Brief. L. Li*. www.ghub.org/cfc_en/wp-content/uploads/sites/2/2011/04/s-Twelfth-Five-Year-Plan-Success-or-Failure-for-Climate-Change-Efforts.pdf

Greenovation Hub. (2014, April). A review of China's climate policies and actions in 2013. *Annual Brief*. L. Li (Ed.). www.ghub.org/cfc_en/wp-content/uploads/sites/2/2014/04/s-Climate-Policies-and-Actions-in-20131.pdf

He, Z. K. (2011). Analysis of the trend of Chinese government innovation: Quantitative research based on the winning projects of the 5 rounds of the "China Local Government Innovation Award", *Journal of Beijing Administration Institute*, 1(1–8).

Hedstrom, P., & Swedberg, R. (1998). Social mechanisms: an introductory essay. In P. Hedstrom & R. Swedberg (Eds.), *Social Mechanisms: An Analytical Approach to Social Theory*. Cambridge: Cambridge University Press.

Heilmann, S. (2008). Policy experimentation in China's economic rise. *Studies in Comparative International Development*, 43, 1–26.

Heinze, T. (2011). Mechanism-based thinking on policy diffusion: A review of current approaches in political science. *KFG Working Paper Series No. 34*. www.polsoz.fu-berlin.de/en/v/transformeurope/publications/working_paper/wp/wp34/WP_34_Heinze.pdf

Hutchcroft, P. D. (2001). Centralization and decentralization in administration and politics: Assessing territorial dimensions of authority and power. *Governance: An International Journal of Policy and Administration*, 14(1), 1, 23–53.

Jordan, A., & Huitema, D. (2014). Innovations in climate policy: The politics of invention, diffusion, and evaluation. *Environmental Politics*, 23(5), 715–734.

Lazer, D. (2001). Regulatory interdependence and international governance. *Journal of European Public Policy*, 8(3), 474–492.

Liu, W. 刘伟. (2014). 社会嵌入与地方政府创新之可持续性——公共服务创新的比较案例分析 [Social embeddedness and sustainability of local government innovation-a comparative case analysis of public service innovation]. *Nanjing shehui kexue*, 1, 87–93.

Maggetti, M., & Gilardi, F. (2016). Problems (and solutions) in the measurement of policy diffusion mechanisms. *Journal of Public Policy*, 36(1), 87–107.

Meseguer, C. (2006). Rational learning and bounded learning in the diffusion of policy innovations. *Rationality and Society*, 18(1), 35–66.

Messner, D. (2015). Shaping global sustainability in the umbrella of "comprehensive globalization": Germany's role. *The Chinese Journal of Global Governance*, 1(1), 16–35.

Mintrom, M. (1997). Policy entrepreneurs and the diffusion of innovation. *American Journal of Political Science*, 41(3), 738–770.

Qi, Y. 齐晔. (Ed.). (2011). *中国低碳发展报告 2010* [Annual Review of Low Carbon Development in China: 2010]. Beijing: Kexue chubanshe.

Qi, Y., Ma, L., Zhang, H., & Li, H. (2008). Translating a global issue into local priority: China's local government response to climate change. *The Journal of Environment Development*, 17(4), 379–400.

Ran, R. (2013). Perverse incentive structure and policy implementation gap in China's local environmental politics. *Journal of Environmental Policy & Planning*, 15, 17–39.

Schröder, M. (2012). *Local Climate Governance in China: Hybrid Actors and Market Mechanisms*. Basingstoke: Palgrave Macmillan.

Shearman, D., & Smith, J. (2007). *Climate Change Challenge and the Failure of Democracy*. Westport, CT: Praeger Publishers.

Shipan, C. R., & Volden, C. (2008). The mechanisms of policy diffusion. *American Journal of Political Science*, 52(4), 840–857.

Shipan, C. R., & Volden, C. (2012). Policy diffusion: Seven lessons for scholars and practitioners. *Public Administration Review*, 72(6), 788–796.

Simmons, B. A., Dobbin, F., & Garrett, G. (2006). Introduction: The international diffusion of liberalism. *International Organization*, 60(4), 781–810.

Simmons, B. A., Dobbin, F., & Garrett, G. (2008a). Introduction: The diffusion of liberalization. In B. Simmons, F. Dobbin, & G. Garrett (Eds.), *The Global Diffusion of Markets and Democracy*. New York: Cambridge University Press, 1–63.

Simmons, B. A., Dobbin, F., & Garrett, G. (2008b). *The Global Diffusion of Markets and Democracy*. New York: Cambridge University Press.

State Council. (2008). *China's Policies and Actions for Addressing Climate Change*. www.ccchina.org.cn/WebSite/CCChina/UpFile/File419.pdf

Teodoro, M. P. (2009). Contingent professionalism: Bureaucratic mobility and the adoption of water conservation rates. *Journal of Public Administration Research and Theory*, 20(2), 437–459.

TWI2050—The World in 2050. (2018). Transformations to achieve the sustainable development goals. *Report prepared by the World in 2050 Initiative*, International Institute for Applied Systems Analysis (IIASA), Laxenburg, Austria. http://pure.iiasa.ac.at/15347

Walker, J. L. (1969). The diffusion of innovations among the American states. *American Political Science Review*, 63(3), 880–899.

Wang, P. H., & Lai, X. J. (2013). 中国公共政策扩散的模式与机制分析 [Analysis on the Modes and Mechanisms of China's Public Policy Diffusion], 6.

Weyland, K. (2005). Theories of policy diffusion: Lessons from Latin American pension reform. *World Politics*, 57(2), 269–295.

Wu, J. N., Ma, L., & Yang, Y. Q. 吴建南, 马亮, 杨宇谦. (2007). 中国地方政府创新的动因、特征与绩效—基于 "中国地方政 府创新奖" 的多案例文本分析 [The motivation, characteristics, and performance of China's local government innovation: Based on the "China Local Government Innovation Award" multi-case text analysis]. *Guanli shijie*, 8, 43–51.

Yang, D. F. 杨代福. (2016). 西方政策创新扩散研究的最新进展 [Latest developments of research on the diffusion of Western policy innovation]. *Guojia xingzheng xueyuan xuebao*, 1. http://theory.people.com.cn/n1/2016/0126/c217905-28086624.html

Yang, H. S. 杨宏山. (2013). 双轨制政策试验: 政策创新的中国经验 [Dual-track policy experiment: Chinese experience of policy innovation]. *Zhongguo xingzheng guanli*, 6.

Yang, X. D. 杨雪冬. (2011). 过去 10 年的中国地方政府改革—基于中国地方政府创新奖的评价 [China's local government reform in the past 10 years: Based on the evaluation of the Local Government Innovation Award]. *Gonggong guanli xuebao*, 1, 81–93, 127.

Yu, K. P. 俞可平. (2010). 应当鼓励和推动什么样的政府创新—对中国地方政府创新奖入围项目的评析 [What kind of government innovation should be encouraged and promoted: Comment and analysis on the shortlisted projects of China Local Government Innovation Award]. *Hebeu xuekan*, 2, 123–128.

Yu, X. H. 于晓虹. (2013). 地方创新的局域性扩散—基于山东新泰"平安协会"实践的考察 [The localized diffusion of local innovation—based on the investigation of Shandong Xintai Safety Association's practice]. 6, 39–43.

Zhang, Z. X. (2015): Carbon emissions trading in China: The evolution from pilots to a nationwide scheme. *Climate Policy*, 15(sup1), S104–S126.

Zhou, W. 周望. (2012). 政策扩散理论与中国政策试验研究: 启示与调适 [Policy diffusion theory and Chinese policy experimental research: Enlightenment and adjustment]. *Sichuan xingzheng xueyuan xuebao*, 4, 43–46.

2 Climate governance and policymaking in China

The "Party State" is a fundamental feature of China's governance structure. China is a single-party state, where the CCP is the absolute power center and has de facto control over all important political decisions (Davidson, 2019). It exercises top-down control through two major systems, namely the cadre evaluation system and the nomenklatura system, both of which are managed by the Organization Department of the Central Committee of the CCP. The former is used to review the cadres' political performance, the result of which influences decisions on promotion (Burns & Wang, 2010). The achievement of targets assigned to lower-level governments by the central government on an annual basis is the main criteria (Turiel et al., 2016). The nomenklatura system is a major instrument of CCP control of China's political, economic, social, and cultural institutions. A list of high-ranking government officials forms the first draft of nominations when a high-ranking position becomes vacant (Leutert, 2018).

Formal state or governmental institutions, through which important policies and measures are developed and administrated, are led by CCP members. The state hierarchy is headed by the State Council and consists of governments at every level, paired with CCP party secretaries or with overlapping CCP and state leadership. This government structure also helps determine non-leadership staffing decisions and control local government agency budgets (Hart et al., 2017). Formally, every government organization across different levels has a bureaucratic rank, which has huge impact on the political weight of mandates.

The Tiao-Kuai system

Policymaking in China is embedded in a vertically and horizontally integrated network of actors. China's quasi-federal arrangement of administration is the *Tiao-Kuai* system. *Tiao* is the set of functional, vertical relationships along different levels of government; *Kuai* is the territorial, horizontal relationships across government agencies at the same administrative level. Together, these

DOI: 10.4324/9781003325307-2

determine to whom a given agency reports (Lieberthal & Oksenberg, 1988; Schurmann, 1968). Multilevel governance, incorporating the vertical (many levels) and horizontal (many actors) dispersion of central government authority (Bache & Flinders, 2004), is a highly relevant normative concept. An overview of the governance ecosystem follows.

1 Central government: top-down institutional or policy change. The power center at the top is the advocate and organizer of the policy reform, relying on administrative orders, legal norms, and etiquette to stimulate the planning, organization, and implementation of institutional innovation in a pyramid-shaped administrative system. Compared to the federal system, a unitary state's higher-level government has greater influence on the lower-level government. China is a typical unitary state with a five-level vertical administrative system composed of (a) the central government, (b) provincial, municipal, and autonomous regions, (c) prefecture-level cities, (d) counties, and (e) townships (see Figure 2.1).
2 Lower-level government: according to new institutional economics, induced institutional change is the change/substitution of the current institutional arrangement, or the creation of a new one, in which a person or group of people respond to opportunities, initiating and implementing policy reforms. The reform process is bottom-up. In a country as vast as

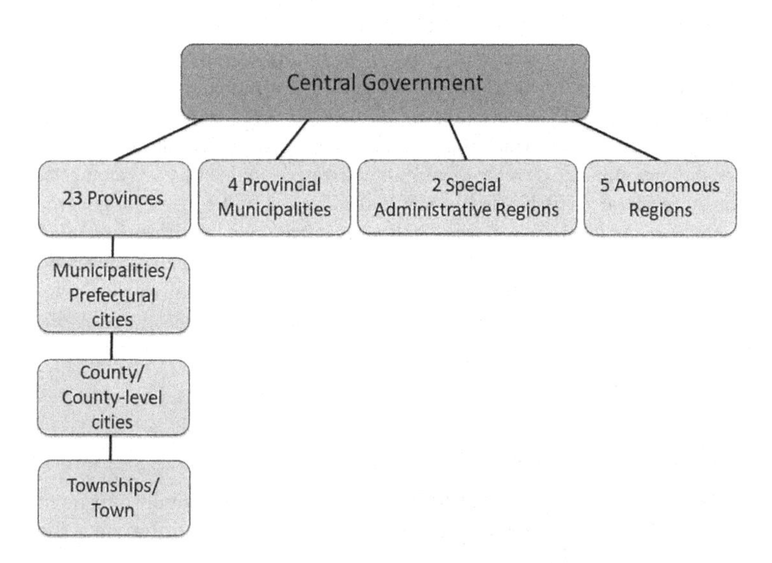

Figure 2.1 China's five-level vertical government administrative system.

Source: Author

China, subnational governments have considerable space for policy exploration, through which they demonstrate their initiative.

3 Neighboring government: geographically adjacent regions may face similar economic and social issues and share compatible social environments. They may also communicate with and learn from each other easily. Greater public pressure and competition may also arise from neighboring governments' policies and facilitate policy diffusion.
4 Concept promotion/norm-leading region: a region that frequently and informally influences the behavior and attitude of other regions, passively or proactively impacting how other regions introduce a new policy.
5 Government with oblique relationship: governments may have oblique connections through which they influence each other (D. F. Yang, 2016). This is the relationship between pluralistic local governments with different administrative levels and without direct governing relations. This could be for example between province A and city C in province B.
6 National or regional network: policy networks are the formal institutional and informal linkages between governmental and other actors structured around shared beliefs and interests in public policymaking and implementation (Rhodes, 2006). Within these networks, some actors, for example, are members of knowledge- or idea-based discourse or epistemic communities. Others actively formulate policy options and serve as members of political coalitions or instrument constituencies (Howlett et al., 2017).

The unit of policy diffusion analysis used here is the broader notion of jurisdictions. The theoretical framework of this book focuses on regional level governments and their role in climate policy diffusion.

Climate policy and ETS in China: from piloting to national policy

After 40 years of rapid economic development, China is now not only the world's second largest economy, but also the largest energy consumer and emitter of greenhouse gases (GHGs) (IEA, 2010; WRI, 2017). It has emphasized its commitment to decoupling emissions from growth, so both domestic and international factors drive the development of a comprehensive policy package to reduce emissions.

China's climate policy began its evolution in the 2000s, but this has accelerated over the past decade. The first of several National Climate Change Assessments over the years was developed between 2002 and 2006 (First National Climate Change Assessment Report Editorial Committee, 2006). The National Climate Change Program (NCCP) was released in 2007 (State Council, 2007), outlining objectives and specific policies to address climate change up to 2010. It was China's first thorough climate policy document and one of the first of its kind in developing countries. Since 2008, China has released strategic climate policy white papers on an annual basis (State Council, 2008).

China enshrined several 2030 climate targets in its first nationally determined contribution (NDC) in 2014 (NDRC, 2014).[1] These include peaking CO_2 emissions around 2030; lowering CO_2 emissions per unit of GDP by 60%–65% from 2005 levels; and increasing the share of non-fossil fuels in primary energy consumption to around 20%. These targets are implemented through a regulatory framework that includes the FYPs (Table 2.1). The 13th FYP further allocates these targets to individual provinces.

The challenge of delivering on the FYP targets has inspired the government to shift from a command-and-control approach towards exploring market measures. Specifically, China has endorsed emissions trading as a vital market-based instrument for climate mitigation.

In an ETS, a regulator defines an upper limit (**cap**) on GHG emissions that may be emitted in clearly defined sectors of an economy (**scope**). Emission permits, or "allowances", are given out or auctioned to entities covered by the ETS (**allocation**). The emissions of covered entities are monitored, reported, and verified (**MRV**). By the end of a defined period, the covered entities must surrender allowances corresponding to their emissions during

Table 2.1 China's climate and energy targets

Target	*2015 (12th FYP)*	*2020 (13th FYP)*	*2030 (2014 NDC)*
Energy intensity	16% reduction compared to 2010 (achieved 8.2% reduction)	16% reduction compared to 2015	
Carbon intensity	17% reduction comparing to 2010 (achieved 20% reduction)	18% reduction compared to 2015, 40%–45% reduction compared to 2005 (Copenhagen Pledge)	60%–65% reduction compared to 2005
Non-fossil fuel share of total energy consumption	11.4% (achieved 12%)	At least 15%	At least 20%
Total energy consumption	4 billion tons coal equivalent (tce)	5 billion tce	6 billion tce
Total GHG emissions	n/a	n/a	Peak CO_2 emissions by 2030

Source: Author, based on China (2016a, 2016b)

that period (**compliance**). Those that have emitted less than the number of allowances they hold can sell any excess to other market participants, while those facing a deficit can comply by purchasing allowances from the market (PMR & ICAP, 2021). Allowing companies to determine when and where to reduce their emissions in this way makes ETS a flexible and cost-efficient policy instrument (ibid.).

To understand how ETS has become part of China's climate policy framework, we must examine Chinese tradition in policymaking: policy experimentation. Here, the central government selects regions in which to experiment, where they are freer regarding policy design. It then expands this new policy geographically or promotes it as a national policy based on the feedback and experiences gained in the pilots, relying in particular on "model" experiences from pilots deemed most successful.

ETS policy development in China follows such an approach. It builds on positive experiences with the Clean Development Mechanism (CDM), a flexibility mechanism under the Kyoto Protocol allowing countries to fund GHG reduction projects in other countries and claim the saved emissions as part of their own efforts to meet international emissions targets. It is further motivated by the advantages of ETS such as cost-effective mitigation, co-benefits like air pollution control, and supporting ecological efforts via the use of offsets.

In October 2011, China announced that it had approved ETS pilots in seven geographically and socioeconomically diverse regions. The aim was to test the feasibility of the market-based policy and facilitate the exploration of its design in various contexts. These include all four provincial municipalities (Beijing, Shanghai, Tianjin, and Chongqing), two provinces (Guangdong and Hubei), and one special economic zone, which is also the only sub-provincial level pilot (Shenzhen, a city in Guangdong province).[2] Geographically, two are in the central and western regions, while five are in the east (see Figure 2.2).

These regions have different economic and energy characteristics. The level of development ranges from slightly below the national average (Chongqing and Hubei) to well above the average (Shenzhen, Shanghai, Beijing, and Tianjin). The varying size of the industrial sector and as such their levels of industrialization and urbanization also mean that these regions vary in the amount of energy they consume. Most are also located in China's major air pollution control zones: the Jing-Jin-Ji Area (Beijing and Tianjin), the Yangtze River Delta (Shanghai), and the Pearl River Delta (Guangdong and Shenzhen). Hubei and Chongqing also suffer from worsening air pollution (J.J. Zhang et al., 2017).

After intensive preparations, the ETS pilots began operation in 2013–2014. The Shenzhen ETS—the first—began operation on 18 June 2013. Other pilots launched soon after.

China's ETS policy development process is marked by another milestone: the political launch of the national ETS in December 2017. This was a goal set in 2015 at the country's highest political level. The provisions for this launch

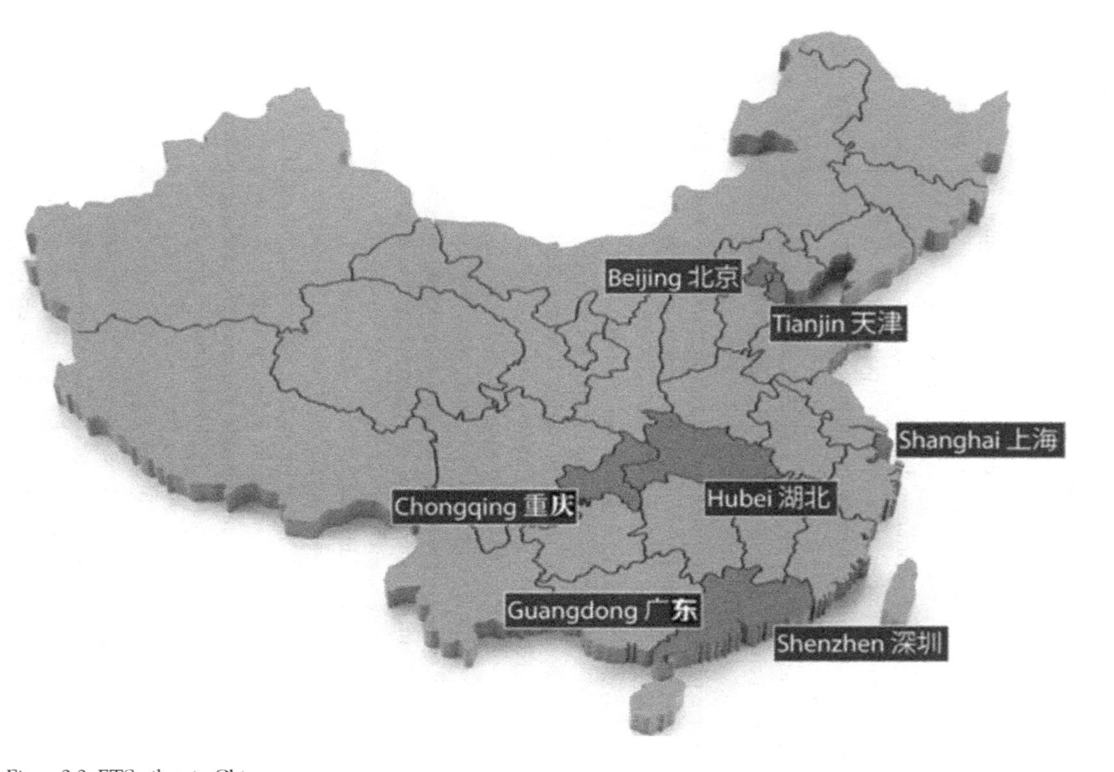

Figure 2.2 ETS pilots in China.

Source: Swartz (2016)

Table 2.2 Basic economic and energy characteristics of the seven pilots (2014)

Region	*Population (millions)*	*GDP per capita (RMB/ person)*	*Secondary industry (%)*	*Energy consumption (tce/person)*	*Energy intensity (tce/10^4 RMB)*
Beijing	21.52	99.16	21.31	3.17	0.36
Tianjin	15.17	103.72	49.16	5.37	0.58
Shanghai	24.26	97.19	34.66	4.57	0.53
Guangdong	107.24	63.26	46.34	2.76	0.49
Shenzhen	10.78	149.56	42.57	3.44	0.23
Hubei	58.16	47.09	46.94	2.81	0.67
Chongqing	29.91	47.70	45.78	2.87	0.68
National	*1362.46*	*50.25*	*46.83*	*3.23*	*0.72*

Source: J.J. Zhang et al. (2017)

Note: Data from 2014, the first year that all seven pilots were operationalized.

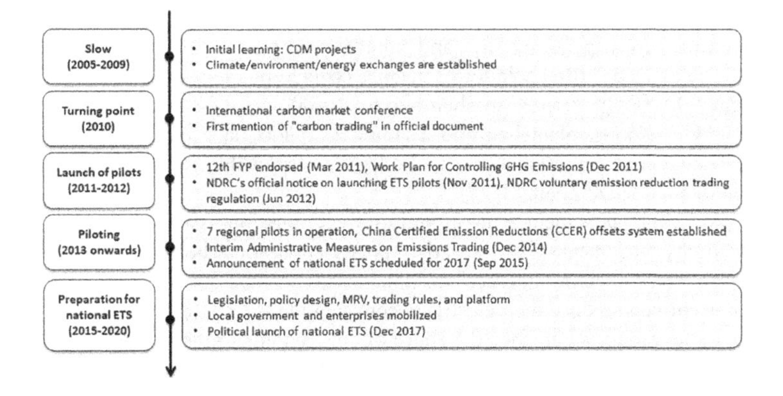

Figure 2.3 Evolving carbon market policy.

Source: Author

and incremental development of the ETS were laid out in the Work Plan for Construction of the National ETS (Power Sector), which was approved by the State Council late in 2017. Figure 2.3 provides an overview of this policy process.

This book divides China's ETS policy development process into two phases:

- Phase One: 2011–2015 (announcement of ETS pilots to announcement of the plan to launch the national ETS); corresponds to the 12th FYP;
- Phase Two: 2016 onwards (in practice, due to the period in which interviews were conducted, phase two covers 2016–2019); corresponds to the 13th FYP.

We examine the dominant policy diffusion mechanisms in each of the two phases for the selected subnational regions, paying attention to any differences across phases.

The ETS pilots

The pilot regions have been flexible in determining their own market rules, following some general guidance from and regularly reporting to national superiors during the design phase (J. J. Zhang et al., 2017). Together, the regional pilots covered 1.2 billion tCO_2 per year, about 11.4% of national emissions in 2014, the first year that all pilots were up and running (ibid.).

All pilots, except Chongqing, cover only **CO_2 emissions**, primarily to simplify MRV requirements. **Sectoral coverage** reflects the regions' heterogeneous economic and industrial structures. At the outset, the electricity, manufacturing, and energy-intensive industrial sectors were covered in all pilot jurisdictions, while the aviation sector only in Shanghai and Guangdong. Shanghai also covers fewer common sectors like shipping and railways, and Shenzhen covers ports, subways, and public buses.

The emissions **threshold** for companies covered by the pilot ETSs ranges from 3,000 tCO_2e per year in Shenzhen to 20,000 tCO_2e per year in Chongqing, Guangdong, Shanghai, and Tianjin. All pilots have enterprises, not individual emitting installations, as the **legally responsible entities for ETS compliance** (Z. X. Zhang, 2015).

All of the pilots include **direct and indirect emissions**. China's regulated electricity price means that the carbon price signal cannot be passed to electricity users. Covering indirect emissions ensures that industrial users have an incentive to reduce their electricity consumption (D. Zhang et al., 2014).

These variables influence the **share of covered emissions in total emissions** in each pilot (35%–60%) and the **number of entities covered** (over 100 entities in Tianjin to more than 800 in Shenzhen).

Cap-setting approaches also differ, though they are mainly bottom-up, aggregating the emissions reduction potential for the covered sectors, subsectors, or participants (Pang & Duan, 2016). The pilot caps are flexible, with only a few predetermined (ibid.).

There are three primary ways to **distribute allowances** in an ETS: (1) free allowances via grandparenting (based on historical emissions or intensities), (2) benchmarking (based on performance indicators), and (3) auctioning. Despite differing approaches and formulas for allocation, the regional ETS pilots have thus far been allocating almost all allowances for free, with the exception that Guangdong auctions a small share. Ad hoc auctioning has been conducted by Shanghai, Hubei, and Shenzhen, largely to offer compliance entities more allowances and/or enhance market liquidity and facilitate price discovery. Most regional ETS pilots use both free allocation methods.

For existing entities, grandparenting is the most common allocation method, while benchmarking is being introduced or developed in several pilots.

To lower the cost of compliance, the pilots have **flexibility mechanisms** in place. All allow entities to "**bank**" allowances to use in future compliance periods,[3] though they forbid "**borrowing**" due to concerns regarding the negative effect on market liquidity and price discovery (Pang & Duan, 2016). Domestic offsets are allowed in all regional ETS pilots for the covered entities to partially meet their abatement. All accept credits from the national CDM-style offset program, the Chinese Certified Emission Reductions (CCERs), while some also have their own domestic programs. Quantitative restrictions on CCERs also apply, based on project type, date of origin, or location.

Each pilot has established an **MRV** system. Local MRV guidelines are set at the sectoral or institutional level. Verification costs are covered by the local government in most pilots, who also select the verifiers (J. J. Zhang et al., 2017). All pilots have a **compliance system**, with penalties ranging from naming-and-shaming and financial payments to future ineligibility for government support. All except Tianjin imposed financial penalties from the very beginning, either a set amount or linked to the carbon price.

The pilots are still young and have generated only moderate trading activities with limited impacts on mitigation and cost-saving (M.Y. Zhang et al., 2017). However, their primary goal has been to allow China to experiment with ETS as a policy instrument. Donning this lens, they have been very successful in reinforcing China's capacity and confidence in the use of market-based measures. For instance, systems have a very high compliance rate—above 99% in most cases (J. J. Zhang et al., 2017).

Allowance prices in most markets rose during the early stages of trading and then declined gradually. Most transactions occur in the period that leads up to the compliance deadline. All allowances are transacted in spot markets; carbon futures have not yet been introduced.

Institutional arrangements for climate policy in China

Based on the Party State and the *Tiao-Kuai* structure, the institutional setup for climate policy in China can be viewed on three levels: legislative and framework, programmatic, and provincial (Fekete et al., 2013). Since ETS piloting began, climate change was for the greater period under the remit of the National Development and Reform Commission (NDRC) at the central level and the respective provincial DRCs. With institutional reform introduced in 2018, such responsibility shifted to the newly established mega Ministry of Environment and Ecology (MEE).

At the **legislative level**, the Chinese Constitution states that the National People's Congress (NPC) and its Standing Committee may exercise legislative power of the state, amending existing and introducing new laws (Government of China, 2014). The NPC issues decisions on climate change mitigation, giving the mandate to the government to integrate mitigation targets into the FYPs. Its endorsement of climate targets also formally binds them at the national level. Energy saving and GHG mitigation are very interlinked in China's climate administration.

At the **framework level**, the State Council leads the executive branch of the government where ministries and state commissions are represented. It also oversees the work of governments in each of China's 34 provincial administrations. Decisions and orders of the State Council, most notably the FYPs, provide the basic framework for all important economic and social development policies and measures, including on climate and energy. As a ministry-level agency directly under the State Council and the macroeconomic planning body of the country, the NDRC is one of the most politically powerful agencies in the central government, with the formal authority to issue instructions to other ministries (Lieberthal & Oksenberg, 1988). The NDRC also administrates the National Energy Administration (NEA). The State Council itself also oversees other agencies related to broader climate policy, such as the China Meteorological Administration (CMA) and the National Bureau of Statistics (NBS) (Government of China, 2018).

At the **programmatic level**, climate policy was under the responsibility of the CMA until the late 1990s (China, 2012; Held et al., 2011), but this was transferred to the NDRC in 1998, shifting climate change from a scientific to a development issue. In 1990, the National Coordination Committee on Climate Change (NCCC) was established under the former State Council Environmental Protection Committee. In 2007, the State Council established the National Leading Group on Climate Change (NLGCC) headed by the premier and with more than 20 ministries and agencies as members. It is the highest administrative body overseeing mitigation action, responsible for China's climate policies, providing guidance for central and local governments' response to climate change, and organizing international negotiations (ibid.). In 2008, the Climate Change Department was created under the NDRC (ibid.). A National Climate Change Expert Committee was also established to support decision-making.

Based on the *Kuai* divisions within the *Tiao-Kuai* system, relevant ministries and government agencies are responsible for specific sectoral climate regulations or overarching themes such as budget or data (see Table 2.3).

Research institutions and think tanks are also important policy advisors for the Chinese government. Key national institutions include the Energy Research Institute (ERI) and the National Center for Climate Change Strategy and International Cooperation (NCSC), both formally affiliated to the government ministry responsible for climate and energy (the NDRC until 2018, after which the MEE).

At the **provincial level**, a specific climate change governance framework was set up in 2007, when the State Council requested provincial governments

Table 2.3 Overview of government institutions and climate policy responsibilities (until 2018)

Level	*Government institution*	*Roles and responsibilities*
Legislative and framework	National People's Congress State Council	Highest legislative body Highest executive and administrative body
Programmatic	NLGCC	Highest administrative body overseeing climate policy
	NDRC: CCD National Energy Administration Ministry of Industry and Information Technology Ministry of Finance Ministry of Housing and Urban-Rural Development Ministry of Transportation MOST NBS	Coordinating climate and energy strategies Sectors: energy, industry, IT sector, buildings, transport State budget and financial policy Science and technology Data and statistics
Provincial	Local leading groups on climate change	Overseeing climate policy
	Local DRC	Climate and energy policy
	Regional subsidiaries of the MIIT, MoF, MOST, etc.	Developing and implementing sectoral/supporting policies at the regional level

Source: Fekete et al. (2013)

to establish climate change coordination mechanisms and leading groups. Led by local DRCs until 2018, jurisdictions shifted to regional bureaus of environment and ecology, following national institutional reform. In addition, the *Tiao*-based subsidiaries of the national ministries at the provincial level are responsible for policymaking and implementation for the respective sector or thematic area under their mandate.

Legal and institutional framework for the ETS pilots

Global experience has shown that an integrated legal and regulatory system is necessary for a well-functioning ETS. This influences environmental integrity, cost-effectiveness, reliability, and the credibility of the ETS (OECD & IEA, 2012). Accordingly, the regional pilots have all invested significant efforts into establishing such legal and regulatory bases for their carbon markets. At the national level, the Interim Measures for the Administration of Carbon Emission Trading, a regulation published by the NDRC in 2014, provides basic principles, general provisions related to governance and design elements such as allowance allocation, trading, MRV, compliance, and market oversight. Although it is an administrative rule rather than a national law, it provides regulatory support for the ETS pilots.

Other regional policy instruments that regulate the individual pilots include regulations on carbon trading, offsets, registries, MRV, allowance allocation, and enforcement. To put these into the Chinese legislative and regulatory context, it is important to understand the hierarchy of the Chinese system (see Figure 2.4). The 2014 NDRC Interim Measures belongs to the administrative rules, and the pilots' top-level regulations are either local decrees as in Beijing and Shenzhen or local government rules as in the rest of the pilots.

ETS is a more complicated policy to administrate than traditional command-and-control measures or other market-based instruments like a carbon tax. Careful policy design and administration of market infrastructure are critical and entail a plethora of actors (see Figure 2.5). Although exact setups may vary, ETS requires a robust institutional framework composed of several common actors.

The regulatory body establishes the legislation to provide a firm legal basis for ETS. It also develops complementary regulations specifying technical elements and market rules. The competent authority is the government body responsible for the design and operation of the ETS. It oversees market participants, emission report verifiers, market operators, and the overall functioning of the carbon market. Other sector-specific government bodies may support the competent authority in ETS rulemaking and/or implementation, helping to manage any financial or market risk that comes about due to the range of financial products allowed in many carbon markets. Typically, the competent

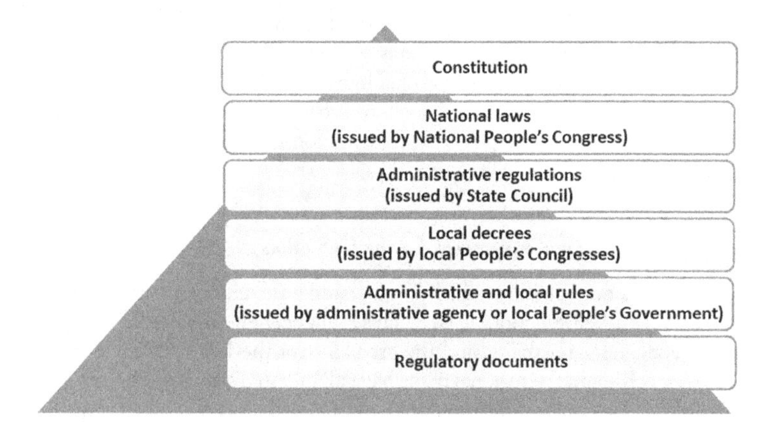

Figure 2.4 Hierarchy of China's legislation system.

Source: Author, based on Legislation of the People's Republic of China

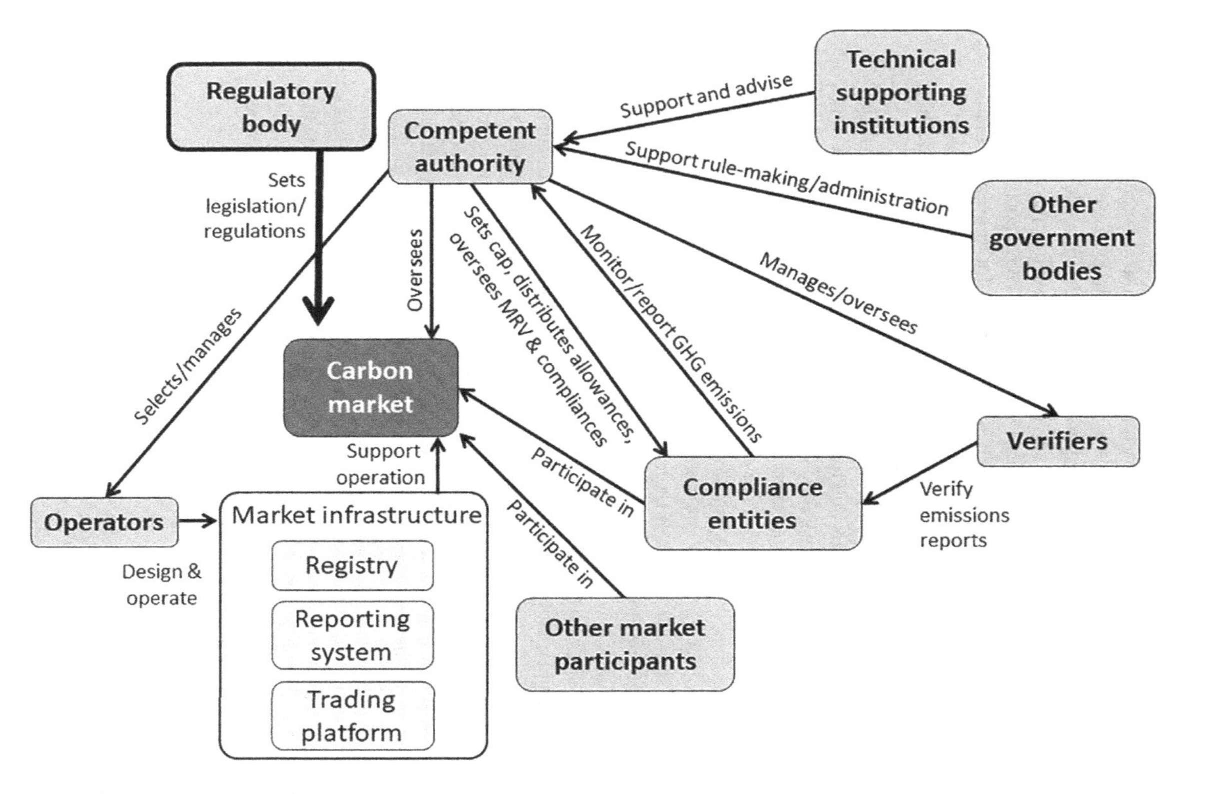

Figure 2.5 Institutions in an ETS.

Source: Author

authority also relies on technical supporting institutions to research and advise on important issues.

Compliance entities or those carrying out specific GHG emissions responsibilities under an ETS are mandatory market participants. In addition, other participants who can voluntarily trade in the carbon market may exist, such as offset project developers, institutional investors, and even individuals.

The market also needs various service institutions. Independent third-party verifiers are accountable for the accuracy of GHG reports submitted as part of MRV. Compliance entities may need support for or even outsource emissions reporting, compliance, carbon asset management, or the identification and implementation of emissions reduction measures. Consultancies, financial institutes, and brokers often provide such services.

A functioning carbon market also requires robust market infrastructure, with systems such as a registry and trading platform essential for market transaction, oversight, and transparency. Qualified institutes may design and operate these systems, under the supervision of the competent authority. Other institutions may be involved in the design and operation of an offset mechanism, or ETS compliance and enforcement measures.

In all pilots, a climate or emissions trading leading group provides political guidance. Often led by a high-level regional government official, it involves the competent authority and other relevant government departments. Some regions set up a new leading group dedicated to ETS, while others defer to existing climate change leading groups or expert committees.

For example, Hubei province set up a leading group to deal with climate change and ETS pilot development and established a committee of experts on carbon trading composed of multi-departmental and interdisciplinary experts, to support the decision-making related to is ETS pilot (China Environment News, 2019). Shanghai established a municipal carbon trading pilot work leading group, led by the municipal leadership, and with the participation of regional authorities. Shanghai also established a Municipality Carbon Emissions Trading Expert Committee with national and local experts and industry representatives to provide guidance, technical support, and decision-making consultation (Shanghai Municipal Government, 2014). Beijing set up a leading group on climate change, headed by the mayor of the municipality, and a new research center to address climate change (Sina Finance, 2018).

Why Hubei and Shanghai?

Hubei and Shanghai stand out among the seven ETS pilots,[4] due to their varied policy influences across China. They have also been pivotal in shaping the national ETS. Interestingly, these two regions have quite different geographic and socioeconomic contexts (see Table 2.2). As a mega municipality and one of the wealthiest cities of the country, Shanghai is on the highly developed eastern coast, one of China's financial and trade centers. In contrast, Hubei, a landlocked province, is in the less-developed Central China region, and its economy quite heavily relies on the industrial sector.

Notes

1 The NDC was updated in 2021.
2 The only overlap between provincial and city pilots.
3 Hubei requires surplus allowances not used for trading in the current compliance period to be cancelled.
4 Based on a survey conducted for this research.

References

Bache, I., & Flinders, M. (2004). Themes and issues in multi-level governance. In I. Bache & M. Flinders (Eds.), *Multi-Level Governance*. Oxford and New York: Oxford University Press, 1–11, 195–206.

Burns, J. P., & Wang, X. Q. (2010, June). Civil service reform in China: Explaining civil servant behaviors. *The China Quarterly*, (212).

China. (2012). The second national on climate change of the People's Republic of China. *UNFCCC*. https://unfccc.int/resource/docs/natc/chnnc2e.pdf

China. (2016a). The 13th five-year plan for economic and social development of the People's Republic of China (2016–2020). *NDRC*. https://en.ndrc.gov.cn/policyrelease_8233/201612/P020191101482242850325.pdf

China. (2016b): 13th five-year plan for renewable energy development. *China Energy Portal*. https://chinaenergyportal.org/en/13th-fyp-development-plan-renewable-energy/

China Environment News. (2019, February 13). 湖北：推进碳市场建设和碳排放交易 [Hubei: Promote Carbon Emissions Trading and Carbon Market Construction]. www.cenews.com.cn/opinion/201902/t20190213_893253.html

Davidson, M. (2019, December). Creating subnational climate institutions in China. *Discussion Paper*. Cambridge, MA: Harvard Project on Climate Agreements.

Fekete, H., Mersmann, F., & Vieweg, M. (2013). *Mitigation of Climate Change in Emerging Countries: From Potentials to Actions*. Dessau-Roßlau, Germany: Federal Environment Agency, 35–57.

First National Assessment Report on Climate Change Editorial Committee. (2006). *First National Assessment Report on Climate Change*. Beijing: Science Press.

Government of China. (2014, August 25). *China's Legislative System*. http://english.www.gov.cn/archive/china_abc/2014/08/23/content_281474982987230.htm

Government of China. (2018, March 17). *The State Council*. http://english.www.gov.cn/archive/china_abc/2014/08/23/content_281474982987314.htm

Hart, C., Zhu, J., & Ying, J. (2017). Mapping China's climate policies. *Embassy of the Federal Republic of Germany*. Beijing. www.chinacarbon.info/sdm_downloads/mapping-chinas-climate-policies

Held, D., Nag, E. M., & Roger, C. (2011). The governance of climate change in China. *Preliminary Report* (LSE-AFD Climate Governance Programme. LSE Global Governance Working Paper 01/2011). www.lse.ac.uk/globalGovernance/publications/workingPapers/climateChangeInChina.pdf

Howlett, M., Mukherjee, I., & Koppenjan, J. (2017). Policy learning and policy networks in theory and practice: The role of policy brokers in the Indonesian biodiesel policy network. *Policy and Society*, 36(2), 233–250.

International Energy Agency. (2010). *China Overtakes the United States to Become World's Largest Energy Consumer*. www.iea.org/newsroomandevents/news/2010/july/name,19716,en.html

Leutert, W. (2018). The political mobility of China's central state-owned enterprise leaders. *The China Quarterly*, 233, 1–21.

Lieberthal, K., & Oksenberg, M. (1988). *Policymaking in China: Leaders, Structures, and Processes*. Princeton: Princeton University Press.

National Development and Reform Commission (NDRC). (2014, August 6). 国家发展改革委关于印发《单位国内生产总值二氧化碳排放降低目标责任考核评估办法》的通知 [Notice of the NDRC on Printing and Distributing the "Measures for Responsibility Assessment and Appraisal of Targets for Reduction of CO2 Emissions Per Unit of GDP"]. www.ccchina.org.cn/nDetail.aspx?newsId=47834&TId=60

Organization for Economic Cooperation and Development (OECD), & International Energy Agency (IEA). (2012). *Making Markets: Unpacking Design and Governance of Carbon Market Mechanisms*. www.oecd-ilibrary.org/environment/making-markets-unpacking-design-and-governance-of-carbon-market-mechanisms_5k43nhks65xs-en

Pang, T., & Duan, M. (2016). Cap setting and allowance allocation in China's emissions trading pilot programs: Special issues and innovative solutions. *Climate Policy*, 16(7), 815–835.

PMR, & ICAP. (2021). *Emissions Trading in Practice: A Handbook on Design and Implementation*. Second Edition. Berlin. https://icapcarbonaction.com/en/publications/emissions-trading-practice-handbook-design-and-implementation-2nd-edition

Rhodes, R. A. W. (2006). Policy network analysis. In M. Moran, M. Rein, & R. E. Goodin (Eds.), *The Oxford Handbook of Public Policy*. Oxford: Oxford University Press, 423–445.

Schurmann, F. (1968). *Ideology and Organization in Communist China*. Berkeley, CA: University of California Press.

Shanghai Municipal Government. (2014, May 13). 上海市人民政府关于本市开展碳排放交易试点工作的实施意见 [Implementation Opinions of the Shanghai Municipal People's Government on the Pilot Work of Carbon Emission Trading in Shanghai]. http://www.cneeex.com/c/2014-05-13/487439.shtml

Sina Finance. (2018, August 16). 北京发改委洪继元: 积极配合建立全国碳交易市场 [Beijing Development and Reform Commission Hong Jiyuan: Actively Cooperate with the Establishment of a National Carbon Trading Market]. http://finance.sina.com.cn/meeting/2018-08-16/doc-ihhvciiv7200942.shtml

State Council. (2007). *National Program on Climate Change*. http://new.fmprc.gov.cn/ce/ceun/eng/chinaandun/economicdevelopment/climatechange/t626117.htm

State Council. (2008). *China's Policies and Actions for Addressing Climate Change*. www.ccchina.org.cn/WebSite/CCChina/UpFile/File419.pdf

Swartz, J. (2016). *China's National Emissions Trading System: Implications for Carbon Markets and Trade*. ICTSD Global Platform on Climate Change, Trade and Sustainable Energy; *Climate Change Architecture Series;* Issue Paper No. 6. International Center for Trade and Sustainable Development; International Emissions Trading

Association; Geneva, Switzerland, 12. https://www.ieta.org/resources/China/Chinas_National_ETS_Implications_for_Carbon_Markets_and_Trade_ICTSD_March2016_Jeff_Swartz.pdf

The First National Assessment Report on Climate Change Editorial Committee. (2006). *First National Assessment Report on Climate Change*. Beijing: Science Press.

Turiel, J., Ding, I., & Liu, J. C.-E. (2016). Environmental governance in China: State, society, and market. *Brill Research Perspectives in Governance and Public Policy in China*, 1(2), 1–67.

World Resource Institute (WRI). (2017). *This Interactive Chart Explains the World's Top 10 Emitters, and How They've Changed*. www.wri.org/blog/2017/04/interactive-chart-explains-worlds-top-10-emitters-and-how-theyve-changed

Yang, D. F. 杨代福. (2016). 西方政策创新扩散研究的最新进展 [Latest developments of research on the diffusion of Western policy innovation]. *Guojia xingzheng xueyuan xuebao*, 1. http://theory.people.com.cn/n1/2016/0126/c217905-28086624.html

Zhang, D., Karplus, V., Cassisa, C., & Zhang, X. L. (2014). Emissions trading in China: Progress and prospects. *Energy Policy*, 75, 9–16.

Zhang, J. J., Wang, Z. X., & Du, X. M. (2017). Lessons learned from China's regional carbon market pilots. *Economics of Energy & Environmental Policy*, 6(2), 1–20.

Zhang, M. Y., Liu, Y., & Su, Y. P. (2017). Comparison of carbon emissions trading schemes in the European Union and China. *Climate*, 5, 70.

Zhang, Z. X. (2015). Carbon emissions trading in China: The evolution from pilots to a nationwide scheme. *Climate Policy*, 15(sup1), S104–S126.

3 Shanghai's ETS and diffusion experience

Perched on the Yangtze River Delta by the East China Sea and with a population of more than 24 million, Shanghai is among the largest and most economically advanced cities in China (Shanghai Information Office & Shanghai Statistics Bureau, 2020). With an annual GDP growth rate of over 6% in the past few years and total GDP of 3.82 trillion RMB in 2019, Shanghai is China's economic, financial, and innovation center (ibid.). The city has a typical post-industrialized economic structure with a high proportion of tertiary industry in its GDP (around 70%) but still has a substantial industrial sector (around 30%) (ibid.; Shanghai Statistics Bureau, 2019a). Important industries in Shanghai include the manufacture of electronic and IT products, automobiles, chemical processing, fine steel products, and biomedicine.

Its per capita GDP was 22,799 USD in 2019, equivalent to that of a high-income country (Shanghai Information Office & Shanghai Statistics Bureau, 2020). Total transactions in Shanghai's financial markets were 1,934.31 trillion RMB in 2019, with the trading volume of several products ranked first in the world. Seven hundred twenty multinational companies have their regional headquarters in the city along with over 460 R&D centers with foreign investment. Shanghai is one of China's most important transport and international shipping hubs. Since 2013, Shanghai has also been a national free trade zone pilot (ibid.).

There is no publicly available data on Shanghai's GHG emissions, and thus its energy consumption is used as a proxy measure. Shanghai's energy structure still heavily relies on fossil fuels, though with a continuous reduction of coal and an increase in natural gas and renewables. In 2015, coal accounted for 35% of the city's primary energy consumption, reduced from the nearly 53% ten years ago, while local renewable energy, imported nuclear, and hydropower together accounted for 13% (Shanghai Municipal Government, 2017b; Zheng et al., 2014). In 2016, fossil fuels accounted for 82% of Shanghai's primary energy consumption, with coal and oil making up 74% and the rest (8%) coming from natural gas (State Grid Shanghai Electric Power Company, 2019; Zheng et al., 2014).

DOI: 10.4324/9781003325307-3

Secondary and tertiary industry account for around 45% and 35% of total energy consumption respectively (Shanghai Statistics Bureau, 2019b). Since the 11th FYP (2005–2010), Shanghai has seen a continuous reduction of energy consumption per GDP. Meanwhile the share of industry consumption as a proportion of final energy consumption dropped from around 60% in 2006 to 50% in 2015 (ibid.). This is partly the result of Shanghai's efforts in energy conservation and emissions control. Unlike some of its ETS pilot peers such as Hubei and Chongqing, who focus their efforts on heavy industrial sectors, Shanghai's low-carbon policy requires a broad-based approach across heavy and light industry, services, and transport.

ETS policy developments

The Shanghai ETS has completed a full policy cycle, from inception through to continuous review and improvement. Unlike some of its pilot peers, the Shanghai ETS has two clear trading periods: 2013–2015 and 2016 onwards.

Alongside its own ETS pilot development (see Figure 3.1), Shanghai is one of the national carbon market training centers and leads the development and operation of the national trading platform.

The Shanghai ETS pilot has a "1+1+N" policy framework: one local government rule, one regulatory document, and a set of supporting documents (see Table 3.1).

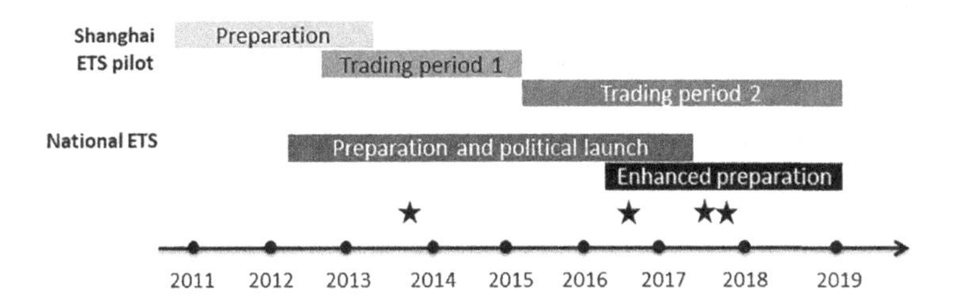

Figure 3.1 Policy development of Shanghai ETS pilot.

Note: ★ *indicates the following milestones:*

Nov. 2013: Shanghai ETS pilot launched

Jul. 2016: National ETS Capacity Building (Shanghai) Center launched

Dec. 2017: Political launch of national ETS

Dec. 2017: Shanghai selected to lead construction of national ETS trading platform

Source: Author

Table 3.1 Shanghai's ETS pilot policy framework

Policy focus	*Document*	*Year published*
General	Trial administrative measures for carbon emissions in Shanghai	2013
	Implementation guidelines for the Shanghai ETS pilot	2012
MRV	Guidance on GHG emissions accounting and reporting in Shanghai (trial)	2012
	Nine sector-specific guidelines on GHG emission accounting and reporting in Shanghai (trial)	2012, waterway transport added in 2016
	Provisional regulations on the administration of ETS allowance registration in Shanghai	2013
	Interim measures for the management of Shanghai's GHG third-party verification agencies	2014
	Trial rules for GHG verification in Shanghai	2014
Coverage	Notice of Shanghai DRC on the list of covered companies in Shanghai ETS	2012, updated annually
Allowance allocation	2013–2015 Allowance allocation and management plan in Shanghai ETS	2013, updated annually since 2016
Trading	SEEE ETS trading rules	2013, amended in 2014 and 2015
	Management for ETS trading membership of the SEEE (trial)	2013, amended in 2014
	Detailed rules for ETS trading settlement of SEEE (trial)	2013
	Management of ETS trading information of SEEE (trial)	2013
	Measures for ETS trading risk control and management of SEEE (trial)	2013, amended in 2015
	Measures for handling violations and breach of ETS trading rules of SEEE (trial)	2013
	Implementation measures for institutional investors in ETS of SEEE (trial)	2014
Offsets	Notice on the use of offset mechanisms in the Shanghai ETS pilot	2015

Source: Author, based on Shanghai DRC (2016)

Shanghai's institutional setup

Shanghai's robust institutions design and manage its ETS (see Figure 3.2). It has evolved over time as a result of changing policy priorities, demand, and the central government's influence.

The Shanghai pilot's 1+1+N regulatory framework has inputs from various levels of government and the main supporting institutions. The Shanghai municipal government, the municipal DRC, and the Shanghai Environment and Energy Exchange (SEEE) share the role of regulator, with the government leading.

Management, characterized by clear division of labor and close cooperation across supporting institutions, was established in 2011. At the top was the newly established Leading Group on Shanghai ETS Pilot Work, responsible

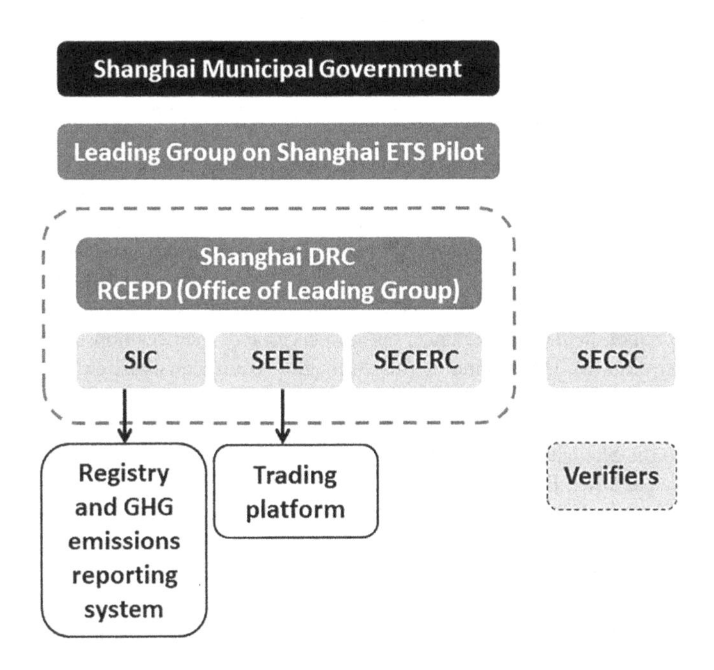

Figure 3.2 Institutional setup of Shanghai ETS (in preparation and early operation phase).

Source: Author based on Shanghai Environment and Energy Exchange (2015); Zhang and Wang (2018); and interviews with Shanghai experts

Note: The larger dashed box indicates that during the early stage of the pilot, the Shanghai DRC and its technical supporting institutions worked in a centralized fashion. Although verifiers are listed as a separate dashed box, both the SIC and the SECERC were part of the first list of ten third-party verifiers selected by the Shanghai DRC in early 2014.

for overall guidance and coordination (Shanghai Municipal Government, 2014). It was led by the executive deputy mayor and composed of responsible heads of the DRC and other municipal departments and agencies, such as the Economic and Information Commission, Finance Bureau, Environmental Protection Bureau, the State-Owned Assets Supervision and Administration Commission (SASAC), Quality Supervision Bureau, Tourism Bureau, and Financial Services Office. The leadership group relied on its office, hosted by the Resource Conservation and Environmental Protection Division (RCEPD) of the DRC, for its day-to-day work. The DRC acted as the overall competent authority of the ETS pilot.

The direct deployment in the office of personnel from the supporting institutions allowed for the exploitation of their knowledge and resources. These institutions include the Shanghai Information Center (SIC), the SEEE, and the Shanghai Energy Conservation and Emission Reduction Center (SECERC). The SIC was responsible for the technical management of the pilot, such as allowance allocation, emission data reporting and verification, and registry system management; the SEEE for developing and operating the trading platform and ETS design; the SECERC provided professional services to industries.

Furthermore, the Shanghai Energy Conservation Supervision Center (SECSC) enforces ETS compliance, where in other pilots this is the competent authority's responsibility. Third-party verifiers validate the quality of emissions data reported by covered entities, helping to ensure environmental and market integrity.

These three technical supporting institutions were selected not only for their carbon market competence, but also because of their relationship with the government. Understanding this thoroughly requires an overview of their institutional histories.

Founded in 1987, the SIC is a government-affiliated institution belonging to the Shanghai municipal government that is directly supervised by the municipal DRC. It is an important decision-making policy advisory body for macroeconomic and social policies and disseminates information to society. The SIC is a member of the National Information Center system, the corresponding institution at the national level directly affiliated to the NDRC. The Center and its subnational centers deal with e-government, data collection and analysis, macroeconomic monitoring and forecasting, policy research, and consulting. The SIC has established a Low-Carbon Economy and Climate Change Center dedicated to policy research and technical support, which provides monitoring, statistics, and certification services.

Established in August 2008 by its parent company, the Shanghai United Property Exchange, and based on the Shanghai municipal government's approval, the SEEE was one of China's first environmental permit trading exchanges. The regional property exchange is the main shareholder or financial investor for the carbon exchange in all pilot regions. The Chinese government in 2008 was far more conservative in its stance on climate policy; GHG trading was officially prohibited, though pollutant permit trading was not.[1] Despite this, Shanghai successfully gained the NDRC's consent to establish the SEEE, which began to trade carbon-related credits, i.e., CDM credits from 2009 and Voluntary Emission Reduction credits from 2011. In 2011, the SEEE was converted into a company limited by shares. Ten shareholders were added, most of whom were national or regional level state-owned companies and institutions.

Approved by the Shanghai DRC and founded in 2010, the SECERC is an SOE belonging to the Shanghai Investment Consulting Company (SICC). Similar to government-affiliated institutions, investment consulting companies (ICCs) in all provinces and provincial municipalities support their regional DRC with decision-making and project evaluation. The SECERC was created from the SICC's former Department of Energy Conservation Consultation. Its mandate was to provide energy-saving services. Its long involvement in this field gave it a good understanding of the relevant industrial companies and their energy consumption. In China, energy-saving and emissions reduction are two sides of the same coin; this gave the SECERC a clear road to work on climate change, including ETS.

The history of these technical supporting institutions and their close ties with the government is clear, especially the Shanghai DRC responsible for the ETS pilot. Their technical competencies "complement each other well" (Interview, SH-E-011420). In conjunction with the time pressure faced by the pilots,[2] that ETS became a complex technocratic project (Lo & Chen, 2019) and contributed to the centralization of institutional arrangements at the early stages of the Shanghai pilot.

The centralized operation of the office was like that of a physical organization (Interview, SH-E-051219), where personnel from the DRC and its three supporting institutions worked together near the DRC. Meetings were held there regularly, where other experts also participated (Interview, SH-E-030220). Centralizing work in this fashion took place only in the Shanghai and Shenzhen pilots (Zhang & Wang, 2018).

Close cooperation between the three supporting institutions and cross-departmental coordination with other relevant governmental agencies were also salient factors during the policy design and early operation stages of the Shanghai ETS. This can be seen for example by how MRV rules were set (see Table 3.2). The SEEE led the drafting of the overarching accounting and reporting guidance, while the SIC and the SECERC drafted sector-specific guidelines. Regardless of which organization led a particular guidance document, the other two supported the process.

Relevant municipal government bodies coordinate cross-departmentally by drafting regulations according to their line of responsibility. The municipal Economic and Information Commission is the policymaker and oversight body for the industrial sector's energy conservation and emissions reduction

Table 3.2 Institutions involved in drafting Shanghai's MRV rules

Policy document	*Institute*	*Participating institutes*
General guidance on GHG emissions accounting and reporting (trial)	SEEE	SIC, SECERC, Statistics Bureau, Economic and Information Commission, Commission of Commerce, Urban and Rural Construction and Transportation Commission, Tourism Bureau, Financial Services Office, Traffic and Port Administration, Bureau of Quality and Technical Supervision
Guidance for electricity and heat generation industry (trial)	SIC	Statistics Bureau, Economic and Information Commission, Shanghai Electric Power Company, Shanghai Electric Power Industry Association, Shanghai Electric Power Co., Ltd., Huaneng Group East China Branch, Shenneng (Group) Co., Ltd., SEEE, SECERC, CQC Shanghai Branch
Guidance for iron and steel industry (trial)	SIC	Baosteel Group Co., Ltd., Statistics Bureau, Economic and Information Commission, SEEE, SECERC
Guidance for chemical industry (trial)	SIC	Shanghai Chemical Industry Association, Statistics Bureau, Economic and Information Commission, SEEE, SECERC
Guidance for nonferrous metals industry (trial)	SIC	Shanghai Nonferrous Metals Industry Association, SEEE, SECERC
Guidance for textile and paper industry (trial)	SIC	SEEE, SECERC
Guidance for nonmetallic mineral products industry (trial)	SIC	Shanghai Construction and Building Materials Industry Market Management Station, Shanghai Building Materials (Group) Corporation, SEEE, SECERC
Guidance for aviation sector (trial)	SECERC	China Eastern Airlines, Shanghai Airlines, China Cargo Airlines Co., Ltd., Shanghai Juneyao Airlines Co., Ltd., Spring Airlines Co., Ltd., Yangtze River Express Airlines Co., Ltd., Urban and Rural Construction and Transportation Commission, East China Regional Administration of Civil Aviation of China, Statistics Bureau, SIC, SEEE

Policy document	*Institute*	*Participating institutes*
Guidance for office buildings (trial)	SECERC	Tourism Bureau, Commercial Commission, Financial Services Office, Statistics Bureau, SIC, SEEE
Guidance for transportation sites	SECERC	Shanghai International Port (Group) Co., Ltd., Shanghai Fubao Port Co., Ltd., Shanghai Airport (Group) Co., Ltd., Shanghai International Airport Co., Ltd., Railway Bureau, Urban and Rural Construction and Transportation Commission, Transportation and Port Management Bureau, Statistics Bureau, SIC, SEEE

Source: Author, based on Shanghai DRC (2016)

policies. It drafted the overarching guidance and three main sector-specific documents. The municipal Statistics Bureau also participated in the development of seven of the ten MRV documents. Similarly, the Bureau of Quality and Technical Supervision was involved based on its responsibility for standard setting.

Shanghai's institutional framework over time

National governance restructuring in 2018 affected all subnational governments in China. Figure 3.3 provides an overview of the Shanghai ETS pilot's institutional setup in its later operation phase. Responsibility for climate change and carbon markets moved from the NDRC to the MEE. This shift also took place at the provincial level, resulting in the Bureau of Ecology and Environment (BEE) of Shanghai becoming the ETS's competent authority. The Division of Atmospheric Environment and Climate Change (DAECC) was set up in the BEE, to take charge of both climate change, including ETS, and air pollution control.

In 2019, the Leading Group on Shanghai ETS Pilot Work merged with the Leading Group on Climate Change, Energy Conservation and Emissions Reduction (Shanghai Municipal Government, 2019). The office of this new leading group was kept in the RCEPD of the DRC, given its role in coordinating energy conservation and as the direct supervisor of the municipal energy bureau.

At the lower level of Shanghai's institutional structure, the distribution of labor and level of the technical supporting institutions' involvement in policymaking have also evolved, driven by changing policy priorities and needs,

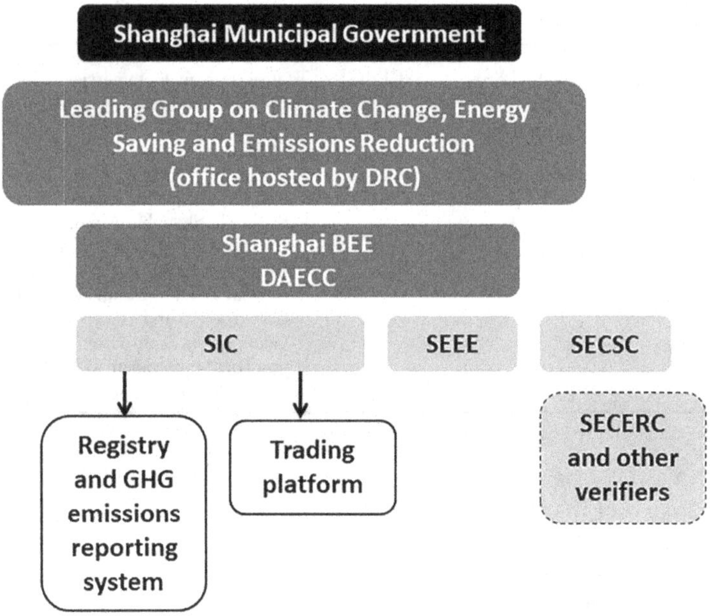

Shanghai Municipal Government

Leading Group on Climate Change, Energy Saving and Emissions Reduction (office hosted by DRC)

Shanghai BEE
DAECC

SIC

SEEE

SECSC

Registry and GHG emissions reporting system

Trading platform

SECERC and other verifiers

Figure 3.3 Institutional setup of Shanghai ETS (in later operation phase).

Source: Author based on interviews with Shanghai experts

Note: The disappearance of the larger dashed box represents that during the later stage of the pilot, the Shanghai DRC and its technical supporting institutions worked in a less centralized mode without a physical joint office. The increased size of the SIC reflects its stronger role in the daily operation of the ETS. The SECERC increasingly moved to acting primarily as a verifier.

institutional interests (e.g., political recognition), and connections (both formal relationships and *Guanxi*). The first change in this regard was the abolishment of the physical office where the DRC and its three supporting institutions worked together in the earlier phase, as focus moved from intensive policy planning to more regular policy implementation.

The SIC has taken an increasingly central role in supporting the government in improving ETS design and managing its day-to-day operations, acting as the technical arm of the DRC (and later the BEE). Since the beginning of preparations for the pilot, the SIC has been responsible for various design elements and market infrastructure. The former, specifically MRV and allowance allocation, were relatively stable during the first trading period, but allocation plans have been tweaked annually during the second trading period. The SIC has led the development of these annual allocation plans, tasked also with the carbon market's daily management, oversight of MRV, and regular communication with covered entities. The SIC has also continued to operate two of the three supporting market infrastructure systems.

In contrast, the SECERC's role, previously policy advisory, now concentrates on providing professional verification services to covered entities. During the pilot preparation phase, its main tasks were to identify which companies were to be covered, develop measuring and reporting methodologies for nonindustrial sectors such as transport and buildings, and collecting historical data (from 2009–2011) for covered entities to provide the basis for their allowance allocation. Once the pilot entered the implementation phase, the SECERC's focus moved to verifying emissions, calculating annual caps, and assessing allowances that were allocated to new plants opened by existing companies. At the end of the first trading period, the Shanghai pilot decided to develop additional rules for new sectors (e.g., waterway transport) in the second period, for which the SECERC was responsible. These duties were similar: developing measurement and reporting rules, collecting historical data, and reviewing allocation methods. The SECERC's tasks were closely linked to its experience; for example, its familiarity with companies' energy usage meant it was well positioned to define ETS coverage. It currently provides third-party verification for the Shanghai pilot.

Like its two peer institutions, the SEEE is tasked with ETS design and operation. Its focus on the market side of the pilot has been consistent throughout phases, supporting the development of regulations and rules for the construction and operation of market infrastructure, developing market products, overseeing market actors, and managing risk. Compared to the SIC and SECERC, the SEEE was more involved in developing the national ETS and hosting the National ETS Capacity Building Center (Shanghai) and was responsible for constructing the national trading platform.

Engaging with the private sector

Unlike in the West, where companies help shape policy via proactive influencing strategies (lobbying) and established consultation processes usually clearly enshrined in law, in China this is much more limited. This is of particular importance for ETS. Firstly, an ETS creates carbon abatement costs for covered entities; these firms are thus naturally concerned as to the fair design and implementation of such a policy instrument. Secondly, ETS is complex and technocratic; companies must regularly manage their accounts under various market settings, monitor and report their emissions, cooperate with third-party verifiers, acquire annual allowances, and subsequently surrender them according to their emissions.

Building on several established consultation and communication mechanisms, the Shanghai government and its supporting institutions have introduced channels through which to engage with companies, taking a "delicate" management approach. Two main objectives include preparing and modifying policies, and communication and support.

During Shanghai's policy preparation phase, SOEs and industrial associations were pivotal in supporting rule setting and developing sectoral guidelines. In addition, the Shanghai ETS pilot has had vast coverage from the very beginning, which requires targeted communication and management. The Office of the Leading Group on the Shanghai ETS Pilot Work thus divided its personnel into several groups based on specific industries and ETS elements (Zhang & Wang, 2018). Each person was responsible for communicating with and providing guidance to companies from up to three sectors, covering all stages of the ETS operational cycle and hosting weekly meetings to discuss progress and problems (Interview, SH-E-030220).

In the operational phase, responsibility for daily communication with companies has moved to the SIC, which acts as a regular two-way channel between the government and covered entities. Via assigned contact counterparts, the SIC distributes ETS information to companies and receives feedback and suggestions to then feed into a continuous policy improvement process (Interview, SH-E-011420).

Another venue for receiving feedback from companies is the "Draft with Request for Comments", an established process in Shanghai. As international experience such as the EU ETS has shown, allocation is very contentious as it has direct implications for companies' costs and competitiveness. Before annual allocation plans in Shanghai are finalized, the draft is sent to all covered entities for formal review (Shanghai BEE, 2019). Though comments are taken into consideration, not all suggestions are necessarily adopted (Interview, SH-E-011420).

Physical meetings are also held to collect stakeholder feedback. The Shanghai pilot expanded its coverage at the start of its second phase in 2016, and newly covered companies and sectors needed special attention. The technical supporting institutions visited selected large companies from the new sectors to gain a better understanding of their status quo and gather feedback in a process called "investigation and research", or *Diaoyan*. *Diaoyan* also occurred during the preparation phase, alongside discussion symposiums and meetings to inform policymaking (Zhang &Wang, 2018). Through these various forums, the technical supporting institutions had held exchanges with nearly 200 companies by the launch of the pilot (Li & Hu, 2012).

Training is another means of communicating with companies and giving guidance on their obligations and opportunities under the ETS. Several training sessions for covered companies have targeted different levels of staff. According to Zhang and Wang (2018), within the first four compliance years, more than 5,000 participants had been trained, with the SEEE and SIC more prominent in organizing the sessions. Interactions between verifiers and companies especially in the initial years of the pilot also provided "soft" training (Interview, SH-E-030220).

ETS communication has also taken advantage of existing cross-governmental relationships as well as between government and companies. For example, when the ETS was under DRC oversight, it worked closely with other line departments—another channel through which to interact with companies. Such cooperation is longstanding, with the DRC having worked on energy conservation for and thereby building close relationships with the Economic and Information Commission. This also helped large energy-intensive companies (many now covered by the ETS) communicate with government authorities (Interview, SH-E-030220).

Institutional and personnel stability

Despite the evolution of the institutional setup, the Shanghai pilot has witnessed much institutional and personnel stability. At the top of the institutional structure, Zhou Qiang (Shanghai DRC) has been heading the Office of the Leading Group since its inception, retaining his role after the 2018 restructuring and the merge of the two Leading Groups (Shanghai Municipal Government, 2019). In addition, the division director Ni Qianlong and the deputy division director Ling Yun were directly responsible for the pilot throughout the whole period that it was under the DRC's management.

The three supporting institutions and their leadership teams have also been stable since the initial preparations. As Zhang and Wang (2018) note, this "effectively ensures the continuity of [the policy's] formulation and implementation." Such stability also ensures a robust working relationship based on interpersonal connections.

Financial resources

Financial resources are crucial to policy innovation and development. As one of China's richest regions, with strong regional political support for environment and climate action, Shanghai has established a dedicated Special Fund for Energy Conservation and Emissions Reduction for both the 12th and 13th FYPs. The fund is much larger than are those of other regions, such as Central and Western China. During the 12th FYP period, 11.3 billion RMB was spent via this fund (Shanghai Municipal Government, 2017a), where money stems from the municipal government budget and income from differentiated electricity prices. The office of the Leading Group on Climate Change, Energy Conservation and Emissions Reduction (hosted by DRC) and the municipal Finance Bureau jointly coordinate on and supervise the operation of this fund (Shanghai DRC, 2018), which has supported climate action, research, and capacity building. There is sufficient funding to support not only necessary but also "forward-looking studies" and action (Interview, SH-E-011420).

Policy diffusion mechanisms

The central government's mandate kickstarts ETS piloting in selected regions and is then followed by a complex and dynamic policy diffusion process under the shadow of the central government's hierarchy.

Both horizontal and vertical ETS policy diffusion is prominent in Shanghai, as a natural result of China's unitary system of government and the resultant governance structure and institutional culture. Various horizontal diffusion mechanisms between Shanghai and other regions (pilot or not) have occurred. Vertically, not only have national policies wrought influence from above, but Shanghai has also critically helped shape the national ETS, injecting bottom-up influence. Expert interviews have also revealed a third type of diffusion by which ETS policy triggers development and diffusion of similar policies: "indirect diffusion".

Horizontal diffusion

Learning is the most dominant policy diffusion mechanism between Shanghai and other ETS pilots and non-pilots, with government authorities and their technical supporting institutions the main drivers in the process. A typical form of learning is government-initiated *Diaoyan* research trips, which are usually arranged as a result of a DRC-to-DRC line request (Interview, SH-E-051219). Particularly during the pilots' preparation and early implementation phases, the NDRC regularly convened meetings for pilots to report on their progress and engage with each other (Interview, SH-E-030120).

Capacity building is another important aspect of learning, with the SEE taking center stage by hosting a National ETS Capacity Building Center,[3] which has trained people across China. The most extensive capacity building for external regions occurred during 2016 and 2017, the first two years after the foundation of the Center. Participants from DRCs, companies, industrial associations, and research institutions were coached on topics such as basic carbon market theory, MRV, national ETS policy, Shanghai ETS operation and management, financial carbon products, and corporate carbon asset management (SEEE, 2015; 2016; 2017; 2018; 2019). The SIC, with its heavy focus on MRV, has also played its part, drawing especially on international donor resources such as the UK's China Prosperity Strategic Programme Fund (UK SPF) to build MRV capacity for Shanghai's neighboring regions.

Policy exchange and learning have also occurred in events hosted by Shanghai, peer regions, the national government, foundations, and other stakeholders. For example, the SEEE organized a series of activities including its branded Green and Low Carbon Salons in 2017–2018, centered on issues such as Shanghai's MRV system, market finance innovations, offsetting, allocation, accounting, and the national ETS (SEEE, 2017; 2018; 2019). Information is crucial for the learning process: the internet, social media, and interpersonal contacts.

The Shanghai ETS pilot is largely seen as successful. The learning mechanism is in this way triggered by rational calculations. Shanghai's experience has also echoed previous studies' conclusions that policymakers are interested in the different elements of a policy's success, i.e., its goals, implementation challenges, and political support. Some regions look to Shanghai to learn how its whole policy package (including ETS) contributes to achieving overall energy or climate targets set by the national government. Others focus on practical policy implementation and how to garner and sustain support and cooperation from covered entities (Interview, SH-E-011420; Interview, SH-E-051219).

Several topics are of interest. Firstly, others are curious about how the Shanghai ETS has observed a 100% annual compliance rate since its inception (Interview, SH-E-271119). Secondly, regional governments are interested in how much has been spent on the ETS every year and the impact of this investment. A carbon market can be pricey; verification, personnel, and management costs for both government and covered companies can be high (Interview, SH-E-051219), particularly for Shanghai with its hands-on management approach. Thirdly, other regions are interested in how to develop market infrastructure and the technical details of the direct GHG reporting system, from system automation to how to establish funding (Interview, SH-E-271119). As a market-based policy, trading under an ETS and the availability and type of market products are also of note.

The learning mechanism involving the Shanghai pilot has spanned different policymaking stages, from agenda setting, policy formulation, adoption, and implementation all the way through to review and adjustment.

Mutual learning between Shanghai and other pilots

Learning diffusion between pilots quite often goes two ways. According to publicly available information, in the early phase of the pilots around 2015, Shanghai received *Diaoyan* delegations from Guangdong, Shenzhen, Beijing, and Hubei, and itself sent delegations to Shenzhen, Beijing, and Hubei (SEEE, 2016). Such exchanges have continued throughout the implementation and fine-tuning process of the pilots. Interviews have revealed what the pilots have learned from each other. Between Shanghai and other pilot regions, learning has focused on referencing each other's policy documents or the design of specific ETS elements, especially MRV and allowance allocation. For the exchanges, pilots have shared experiences regarding trading products and market innovation.

Several interviewees believe that Shanghai's comprehensive and early published administrative measures provided an important reference for other pilots' policy documents. ETS can be a highly uniform policy in terms of general composition. Innovation lies in the specifics of each design element (Interview, SH-E-051219). As others were developing similar documents, Shanghai's served as a natural reference.

Shanghai also published a complete policy package of MRV rules early on, including a measuring and reporting guideline, multiple sector-specific rules, and a verification guideline, which also "served as a reference for others" (Interview, SH-E-051219). A Shenzhen technical supporting institution confirmed this, citing Beijing and Shanghai as reference points for their own measuring and reporting systems (Interview, SZ-E-040220). Vice versa, pilots including Shanghai have taken up as good practice Shenzhen's initiation of fourth-party verification to further improve the credibility of emissions reporting (Interview, SH-E-051219).

Shanghai has learned especially from Guangdong and Beijing about allocation, moving accordingly from a three-year cycle to annual allocation. Shanghai also started the benchmarking method for the aviation and power sectors early, but not for industrial sectors. The city deliberated with Guangdong about benchmark- and efficiency-based allocation for industrial sectors, as Guangdong had already implemented this (Interview, SH-E-051219). Beijing's sector-based standard for allocating allowance for new entrants also influenced Shanghai (Interview, SH-E-030220). Learning here has once again been a two-way process. Other pilots have noted that Shanghai's use of a load factor for power generation equipment as a correction factor is a "good idea", as this ratio significantly impacts carbon emissions (Interview, SZ-E-040220).

Shanghai has also pioneered trading products, including the spot forward trading contract, which Hubei and Guangdong later also initiated (Interview, SH-E-051219). "Our advantage in product innovation is that we focus on making rules (instead of just one-off transactions)" (Interview, SH-E-030120). Rules developed by Shanghai for their products provide a solid basis for others. Other pilots also use the idea of "new products" traded elsewhere to counterbalance any internal opposition to them. For example, when carbon asset pledge loans were first introduced, other exchanges could report to their government authorities "saying 'look, Shenzhen has done it so we can too without (excessive) risk'" (Interview, SZ-E-080220).

Shanghai partakes in many exchanges with the Beijing, Guangdong, and Shenzhen pilots, due to the similarity of their systems and contexts (Interview, SH-E-030120). Learning is thus a continuous and dynamic process not limited to just two pilots. "Different pilots have seen a hundred flowers blossom . . . either they come here, or we go there" (Interview, SH-E-011420).

Shanghai's experience confirms the conceptualization of the learning mechanism in that when jurisdictions refer to others' policies, whether they can be applied to their own context is of great importance. Information not only on "what" but on "why" is thus critical. Carefully weighing the benefits and drawbacks of a certain policy leads to a decision regarding policy adoption in the follower region. For example, Shenzhen was the first pilot to allow individual investors, and from early on. Though Shanghai has since opened its market to institutional investors after heated internal debate, it still keeps individuals out of the market (Interview, SH-E-030220). This case of non-adoption shows the learning entails at least a degree of rationality.

Although learning is continuous, its intensity varies across time. In the design and early operation phase (until around 2015), pilots were motivated to learn from each other regarding allocation methods, sectoral coverage, and trading products (Interview, SH-E-030120). By the time Shanghai's ETS design had taken shape but before launch, Shanghai had conducted many *Diaoyan* with its peers (Interview, SH-E-051219). The factors underpinning intensive learning during this phase are similar to those driving the centralization of institutional arrangement: the complexity of designing an ETS from scratch and significant time pressure.

Though international ETSs are another source of policy learning for the Chinese pilots (e.g., the EU ETS and California's system), they are not the focus of this book.[4]

Diffusion between Shanghai and non-pilot regions: ideas to match domestic policy demand

Many non-pilot regions have actively learned from Shanghai's experiences, for which there are two major motivations.

First, China's piloting usually leads to expansion to more regions before a national rollout. Several non-pilot regions initially considered establishing their own ETS pilots (Interview, SH-E-051219; Goron & Cassisa, 2017). ETS was considered an innovative policy and a potential source of political accomplishment (Interview, SH-E-051219). What they aspired to learn from Shanghai was thus regarding general ETS policy frameworks and management (Interview, SH-E-051219).

Then, around 2015–2016, the central government made clear that no further pilots would be established, and the national ETS was to be constructed directly instead. Non-pilots' motivation changed to learning practical measures from pilots to meet the requirements set by the central government. They concentrated on preparing for the national ETS, including aspects like verification, trading platforms, and verifier selection (Interview, SH-E-051219).

Regions initially considering their own carbon markets and thus actively exchanged with Shanghai were Guizhou, Jiangxi, Yunnan, Shaanxi, Gansu, Jiangsu, and Qingdao (Interview, SH-E-051219; Interview, SH-E-030120). Shanghai experts suggest that even more regions were driven by the motivation of wanting to fulfill national ETS requirements from 2015–2016.

Diaoyan also took place between Shanghai and non-pilot regions. One interesting case is the exchange with neighboring Zhejiang province in the Yangtze River Delta region. Shanghai and Zhejiang are both pioneers in market-based environment and energy policy; the latter was home to one of the first Pollution Rights Trading pilots in China and more recently also the first energy consumption quota trading (ECQT) pilot. High marketization and openness have made pilots like Shanghai attractive for Zhejiang (unlike for example Hubei, which is considered to be more under government control) (Interview, ZJ-E-291219). Geographical proximity, similar market characteristics, and institutional setup enable policy learning between the two regions. In Zhejiang, the Zhejiang Economic Information Center (ZEIC) set up a Center for Climate Change and Low Carbon Development in 2012, the first province-level center of its kind in China. Like the SIC, the ZEIC has been the major supporting institution for the province's market-based instruments and other climate policy in Zhejiang (Interview, ZJ-E-291219).

Shanghai initiated a regional workshop on ETS and cooperation in 2014, during which the deputy director of the Zhejiang DRC expressed interest in learning from Shanghai (via *Diaoyan*) about its policy design, how the implementation process was organized, and the main challenges it encountered (Interview, ZJ-E-291219). When Zhejiang was assigned as the national ECQT pilot, it organized another *Diaoyan* to Shanghai, where it was introduced to further relevant institutions there (Interview, SH-E-030120). Finally, the ongoing process of implementing the Yangtze River Delta Integration strategy, a political task stemming from the national leadership of four regions (Shanghai, Zhejiang, Jiangsu, and Anhui), has provided further opportunities to explore cooperation potential on low-carbon policies. Based initially on two market-based policies—first the ETS and then the ECQT—Shanghai and Zhejiang show deep mutual learning and institutional contact (Interview, SH-E-271119).

Shanghai and non-pilots have also engaged in capacity building. Not only have Shanghai's institutions hosted many training activities across China, experts from Shanghai are also frequently invited to speak at sessions organized by the central government, international development agencies, SOEs, and sector associations.

Shanghai's pilot is highly transparent compared to its peers, which further facilitates learning. Its dedicated Shanghai Energy Conservation, Low Carbon and Climate Change Website is a one-stop shop for ETS policy, announcements, and case studies on good corporate carbon strategies. Shanghai has also published a Compilation of Relevant Documents for the Shanghai ETS Pilot (Shanghai DRC, 2016), which provides information for broad audiences who want to understand details of Shanghai's policies. These materials, however, focus solely on the "what'" rather than the "why'" and so are usually combined with other forms of learning.

Institutionalized cooperation also exists between Shanghai and other regions. The SEEE has established cooperation agreements with partner institutions, thanks to its experiences and competencies accumulated through heavy involvement in the Shanghai ETS pilot design and operation. This is also a result of the SEEE's active business development as a profit-driven company. For example, it has signed strategic or market cooperation agreements with the Shanxi Environment and Energy Trading Center, the Jiangsu (Suzhou) Environment and Energy Trading Center, the Inner Mongolia Environment and Energy Exchange, and the Lanzhou Environment and Energy Trading Center (SEEE, 2017, 2018).

Meetings specifically to share experiences are held with both pilot and non-pilot regions. For example, in 2015 the SEEE held a provincial and municipal meeting on carbon trading experiences, inviting more than 20 provincial pilot and non-pilot NDRCs to discuss how to promote carbon trading in their respective jurisdictions (SEEE, 2016).

Though many experts refer to Shanghai as a "leading" or "advanced" region to which others turn to for new policy ideas or practices, this does not result in 1:1 policy **emulation**. Nowhere has adopted ETS policy purely by mimicking Shanghai, nor has any region simply copied Shanghai's ETS policy design. Furthermore, Shanghai is not the only region referred to as an ETS policy-leading region; experts frequently also mention Guangdong, Shenzhen, and Beijing. Shanghai's experience serves merely as a reference for policymakers and technical supporting institutions elsewhere in China seeking ideas and examples related to ETS. This then simply feeds into the policymaking process of the respective region. As such, emulation does not exist in its purest form.

As the national finance, economic, and innovation center, Shanghai is often regarded as a "role model" in areas related to economic and social development. The city has been identified as a "pioneer in national reform" and a "forerunner of innovative development" (State Council, 2017). The Central Committee of the CCP, together with the State Council, also issued the Outline of the Yangtze River Delta Regional Integration Development Plan in 2019, further endorsing Shanghai's position as the "dragon's head" for the region.

China's eastern coast, where Shanghai is located, is considered to have higher levels of governance capacity and successful policy experience. However, how much Shanghai's leading image is applicable to climate policy is complicated. Many subnational regions view climate policy as a burden and worry that it could restrict local businesses' development. Fear of negative impact on regional economies may thus dilute interest in emulating their climate policies.

Learning is sometimes paired with the **socialization** diffusion mechanism, especially in the form of capacity building. The focus of the socialization mechanism is the norm of ETS, while regional policymakers in China face unique practical issues that require pragmatic solutions. For them, without real, on-the-ground effect, it is hard for any policy idea to gain enough interest for cross-regional adoption. For this reason, though interviews have revealed that socialization does exist, it is seen as a secondary mechanism.

The SEEE and SIC disseminate information on what an ETS is, promoting the normative idea (Interview, SH-E-051219). Publicity is thus another crucial form of socialization. As an advocate of ETS, the SEEE touts its advantages (Interview, SH-E-030120).

ETS events and other forums with broader themes are important venues to promote the concept. For example, Shanghai's Green and Low Carbon Development Summit was initiated by the SEEE in 2016 and is designed as a credible forum in the field of climate change and carbon markets (SEEE, 2017). Renowned speakers such as Xie Zhenhua (former vice-chairman of the NDRC), Liu Yanhua (counselor of the State Council, director of the National Expert Committee on Climate Change), and German environmental ministry officials have bolstered the summit's publicity, though it is difficult to evaluate concretely the exact impact of such events.

The combination of piloting, the nomenklatura system, and the cadre evaluation system has led to virtuous **competition** among pilot leadership (Goron & Cassisa, 2017). We add two modifications to the theoretical framework on (1) the degree of influence and (2) the time dimension of the competition mechanism.

Firstly, experts especially from eastern China agree that competition is a relatively weak ETS policy diffusion mechanism (Interview, BJ-E-040220). This competition is also virtuous, which has led to pilots benchmarking themselves against each other (Interview, SH-E-051219).

Secondly, unlike learning, where impact throughout different stages of the policy process has been stable, the competition mechanism only emerges as relevant at specific points. For example, competition between regional pilot governments occurred once the NDRC organized an open selection process to pick the pilots to lead the construction of the registry and the trading platform for the national ETS (EDF & ERI, 2018). Nine regions (seven pilots plus Fujian and Jiangsu) competed (Interview, SH-E-051219). Interestingly, the sense of competition among regional governments differs significantly. Hubei's is particularly strong, considering victory to be a political achievement beneficial for the future promotion of its leaders. Shanghai did not have this view, regarding it merely as "icing on the cake" (Interview, SH-E-271119).

Some competition also exists between the pilots' carbon exchanges, the most striking case of which occurred when the national government assigned National ETS Capacity Building Centers, and exchanges rushed to "stake their claims in new markets" (Interview, SH-E-030120). This can be seen in the amped-up capacity building activities hosted by the Shanghai Center in other regions in 2016 and 2017. Pilots tried to seize market resources in preparation for the future national carbon market (Interview, TJ-E-040220), and the exchanges' behavior was driven by business logic (Interview, BJ-E-090220).

Coercion diffusion takes place in two main forms, both triggered by material consequences. As the conceptualization of this mechanism lays out, incentives may be either positive or negative—carrots or sticks.

Firstly, policy performance appraisals conducted by the central government's energy and climate targets put pressure on regions to formulate policies to help meet them. The material trigger here is the potential negative impact from the results of the evaluation; regional leaders may lose out on the chance to be promoted. Secondly, pilot regions try to use ETS to incentivize cross-jurisdictional or regional cooperation, which may provide economic or policy benefits (the material triggers). The case of Shanghai exhibits both forms, though with very limited effect.

Central government evaluation system

The 12th FYP saw the distribution of national emissions intensity reduction targets broken down to the province level for the first time. The NDRC published measures for the assessment and evaluation of these targets in 2014 based on a trial appraisal the year before (NDRC, 2014). This system was extended (though modified) for the 13th FYP. Regional governments are assessed on an annual basis against an evaluation matrix; in addition, there is a final assessment for the whole FYP. Scores are assigned to different quantitative and qualitative criteria, with a total attainable 100 points. Results are divided into four numerically calibrated tiers: excellent, good, qualified, and unqualified (ibid.).

The impact of this evaluation system is limited. Although evaluation results are supposed to feed into the local government leaders' cadre evaluation system, it is not clear how exactly this should be done. Compared to the previous evaluation system for energy conservation and pollution reduction targets, this system has fewer restrictions for regional governments. With the national Energy Conservation Law and Environmental Protection Law in place, there is a solid legal basis for energy and pollution initiatives. Failing to meet pollution control targets could result in restrictions from the central government for the approval of new projects in the region, but no penalizing measures are foreseen for failing to pass emissions reduction target evaluations (Wang, 2014). Accountability for emissions reduction is "still relatively soft, unlike that for traditional environmental protection" (Interview, SH-E-011420). This is because accountability for energy conservation or pollution reduction is mainly handled by one department. Carbon involves several departments that must work together (Wang, 2014), making attributing overall responsibility to a single entity difficult. This is especially true after 2018 when climate change (and ETS) moved to the ecology and environment departments and bureaus, traditionally weak departments at the regional level that need to coordinate with other more powerful departments like those responsible for SOEs or industries.

Furthermore, work on ETS accounts for a very small portion of the overall scoring. In its very first version, the "target achievement" category with two overarching quantitative targets (i.e., the annual and accumulative emissions reduction targets) accounted for 50 points, making it the most important category. There were two additional categories of tasks and measures, basic work, and capacity building. Ten quantitative and qualitative criteria were included in these two categories, which accounted for another 50 points. Together with other innovative measures, ETS belonged to a fourth "other" category, which was used as a reference but not included in the final score. ETS was one of the least relevant items in this matrix.

This situation has not changed much, despite central government modifications for the 13th FYP. The first category continues to dominate (though with a slightly reduced score of 40 points), and the two additional categories have been merged (with an adjusted total score of 60 points) (N. Chen, 2019). ETS is an individual subcategory worth eight points: the compliance rate for pilot regions accounts for three points, the enterprises' reporting, verification, and allowance allocation for four points, and other market mechanisms for one point.

This evaluation system does generate some pressure for regional governments by sparking interest to learn from Shanghai how to achieve emissions intensity targets or about policy management (Interview, SH-E-030120; SH-E-051219). Like the learning mechanism, this pressure has led to the exchange of experiences and ideas (Interview, SH-E-011420). The evaluation also provides additional centrally organized occasions for regional governments to share challenges they face and any suggestions for the national government (Interview, SH-E-051219). However, it focuses on broader climate and energy policy rather than ETS specifically.

East China regional cooperation

Shortly after its pilot entered the implementation phase, Shanghai made efforts to promote regional cooperation with its peers in East China, one of the country's six greater administrative areas. In July 2014, the Shanghai DRC and SEEE jointly organized a Regional Cooperation Seminar on ETS, inviting government authorities responsible for climate policy from six provinces to Shanghai to discuss regional ETS cooperation (SEEE, 2014). The seminar was chaired by the secretary general of the Shanghai DRC, which demonstrated the Shanghai government's political support for regional cooperation.

Ideas for regional cooperation were also promoted by Shanghai during events involving East China and Yangtze River Delta peers, including proposals to share the Shanghai trading platform and MRV system or jointly develop regional allowance allocation methods (Interview, SH-E-051219). Another proposal was to use CCER as a first step for other provinces to join Shanghai (without the need to establish an ETS) (Interview, SH-E-030220). However, the political launch of the national ETS in 2017 came shortly after and left little space for such regional cooperation.

Shanghai was not alone in its attempt to promote cross-jurisdictional cooperation nor the only region that failed to do so; this was the case for almost all pilots. The conceptualization of the coercion mechanism foresees difficulties in this diffusion process, in particular stemming from opposition from domestic groups in the policy follower region. But Shanghai's experience has shown that difficulties may also result from national policy developments, a pertinent lesson for countries with a more centralized governance structure like China's. As for pace of adoption, the Shanghai case matches the conceptualization well, which sees a relatively low speed of response from the peer regions' governments.

Vertical diffusion

Shanghai has considered its "very task" to be to influence national carbon market policy, i.e., bottom-up diffusion (Interview, SH-E-271119). Its most outstanding contribution is its construction of the trading platform for the national ETS, one of the three pillars for market infrastructure underpinning the architecture of the national system. After being selected by the NDRC to lead this work, Shanghai arranged for a dedicated venue, personnel, and technical resources (SEEE, 2019). It also supported the national government in designing trading rules for the national ETS. These rules are similar to Shanghai's, whose rules, price formation mechanisms, and management measures have worked well to date (Interview, SH-E-051219).

The national ETS's trading platform also needs to be managed. Regular tasks like system operation and maintenance must be delivered through a defined institutional setup. During the political launch of the national ETS, the NDRC signed contracts with nine provinces to jointly develop and maintain the trading system, which will be under the joint supervision of central and regional governments (Chongqing DRC, 2018).[5] Shanghai provided its suggestions on the organizational form, management charter, and shareholder agreements.

A common channel used to transfer experiences from the pilots to national ETS policymaking are the requests for comment. When preparing national ETS regulatory documents, rules, and technical guidelines, the national government often sent drafts to the pilots for their comments (Interview, SH-E-051219). The updated version based on pilot feedback was then sent to all regional governments for further review. Shanghai provided feedback on issues such as reporting and verification (Interview, SH-E-271119). The central government reviewed these suggestions in the context of the heterogeneity of Chinese regions, and potentially simplified policy design at the national level.[6] Like in the horizontal learning mechanism, *not* adopting policy suggestions from regional governments due to their context-specificity points to "rational" learning on the part of the national government.

Another common channel is *Diaoyan*, which is similar to the learning mechanism at the horizontal level. Typically, the technical supporting institutions of the national ETS visited the pilots to discuss their development and operation (Interview, SH-E-011420; Interview, SH-E-030220). The results of such Diaoyan can be traced by the eventual content of national policies. "The framework and supplementary technical documents of the national ETS policy follows similar logic to ours . . . with obvious spillover effect" (Interview, SH-E-271119).

Technical supporting institutions in Shanghai have also used resources from international donors to provide policy recommendations to and engage with policymakers at the national level. Examples include the German development agency Gesellschaft für Internationale Zusammenarbeit (GIZ) and the UK SPF in 2016, with whom the development of the national ETS verification management system and verification rules for the aviation sector, respectively, were discussed.

The Shanghai ETS pilot has also sent experts to Beijing to join the (eventually only short-term) national ETS preparation working group, set up in April 2017 by the NDRC to accelerate progress (EDF & ERI, 2018). Regional experts joined this group on secondment, bringing technical and regional expertise, though their exact influence cannot be assessed. Indeed, several interviews have revealed the limitations of such secondments as a channel for bottom-up influence on national ETS policy, as the central government did not systematically invite experts for their regional expertise but instead to bulk numbers.

The national government has also regularly invited speakers from regional pilots to training sessions, to discuss both their general ETS experience and whatever ETS element the region has excelled in. For example, the SIC often presents Shanghai's MRV experience. The SEEE talks regularly about trading rules and companies' compliance strategies. Though this did not directly influence national ETS policymaking, it helped non-pilot regions' stakeholders better understand the instrument, moving from conceptual to more practical ETS knowledge.

Officials from the central government have acknowledged Shanghai's role in supporting national ETS development, citing its "successful completion" of tasks set by the NDRC. According to Su Wei, former director general of the Department of Climate Change of the NDRC, not only has the Shanghai pilot proved a valuable reference for the national ETS, but it has also helped transfer these practices to non-pilot regions, in doing so facilitating the establishment of a unified national carbon market (SEEE, 2017).

Top-down diffusion from national government in China is far-reaching, unlike in the West, where horizontal diffusion is dominant. This, however, is unsurprising, given China's unitary system of government. More surprising is the diversity of such influence. Not only are there myriad ways that national ETS policy can have direct (positive or negative) influence on regional policy learning and the diffusion process, beyond emissions trading, but other national policies and arrangements also (sometimes decisively) impact the regional level.

National ETS policy

National ETS policies influence how horizontal diffusion works at the regional level, sometimes also directly impacting regional ETS policy design.

The central government has actively pursued measures to promote regional exchange via training workshops and dedicated capacity building centers, which has accelerated horizontal diffusion. Shanghai has rich experience in this regard.

However, some national level policies can also hinder horizontal diffusion, sometimes unintentionally. The initial interest of several provinces and cities in establishing their own ETS was as a means of demonstrating political performance. These regions approached pilots like Shanghai to learn how to design and manage a carbon market, though this learning process ceased when it became clear that a national ETS was to be developed in a centralized manner. The impact of this decision is mixed. Such a top-down approach to the national ETS combined with extensive preparation assigned to the regional governments created new and different motivations for non-pilot regions to learn from the pilots' practical experiences regarding issues such as historical data collection and verification.

Besides interactions between regional governments, national policy also influences regional ETS policy design directly. While some regions aligned their ETS design with national or even draft rules, Shanghai has been comparatively "immune" to this influence from national ETS policy. Several factors explain why this may be the case: differences in economic and industrial structure, requiring tailored ETS policy, alongside Shanghai's confidence in its ETS and its design.

National level policy also influences the market side of ETS policy innovation and diffusion. After climate responsibility shifted to the MEE, the body has been focusing on progressing the national ETS while emphasizing that local innovation should avoid the overfinancialization and derivatization of market products. Due to this national guidance and changing personnel in charge of ETS matters in Shanghai, recent years have seen increasing constraint on carbon market innovation (Interview, SH-E-030120; Interview, GD-E-170120).

National policies and arrangements beyond ETS

A range of broader national policies, concepts, and arrangements beyond ETS also influence horizontal diffusion. These include:

1. Policy documents and guidance issued by the central government, which may increase local governments' motivation to pursue ETS as a source of political performance and thus trigger policy diffusion.
2. Other comprehensive or policy-specific experimentation approved or promoted by the national government, e.g., pilots for free trade zones, green finance, and ecological compensation mechanisms. As earlier, these may contribute to local governments pursuing ETS policy, leading to (in)direct diffusion.

3 The "counterpart help" mechanism, where the central government assigns province pairs under the umbrella of poverty eradication and development policy. This may facilitate broader policy exchange and cooperation between the paired regions, including (in)direct ETS policy diffusion.[7]

The second of the prior points is particularly relevant for Shanghai. The Shanghai-Zhejiang exchange case has shown that environmental and energy rights (outside of emissions) trading policies contribute to policy diffusion, especially when the technical supporting institutions behind them overlap with ETS. However, the coexistence of multiple market-based instruments in energy and climate can also raise the issue of policy coordination, the lack of which can lead to confusion and significant burdens for regional governments. It can also create unnecessary policy competition at the regional level, resulting in a shift away from ETS towards other market-based policies instead.

Indirect diffusion

Interviews with experts in Shanghai have also revealed another type of diffusion by which ETS policy triggers the development and diffusion of other relevant policies. This is a mechanism not foreseen by the theoretical framework this book bases on previous literature, but one that emerged later during interviews: "indirect diffusion".

There are two sub-models of indirect diffusion (see Figure 3.4). In the first, policy A triggers a relevant policy B in the originating region; then, policy diffusion occurs based on B between the originating region (policy supplier) and the new region (policy follower). In the second, policy A is implemented in the policy supplier region, while the policy follower region is either considering or preparing relevant policy B. The regions then engage with each other so that the follower region may advance its policymaking process. Interestingly, while Shanghai displays stronger evidence of the first model, Hubei shows stronger evidence of the second model.

Shanghai's most prominent case of indirect diffusion is that of carbon finance. As China's financial hub, Shanghai has long aimed to transform itself into four essential pillars on a greater scale: the world's finance, trade, shipping, and economic center. In its first ever ETS policy document, the pilot implementation guideline issued by the municipal government in 2014, Shanghai made its ambition clear: to advance innovative carbon finance markets and become the carbon market and trading platform for the national ETS, in order to promote this "four centers" idea (Shanghai DRC, 2016). To realize this, Shanghai has established a new institutional structure and developed a spectrum of carbon finance policies. These then provide a good basis for carbon finance policy learning between Shanghai and other regions.

Supported by the municipal DRC and Financial Services Office,[8] in 2015 Shanghai established a coordination group for carbon financial market

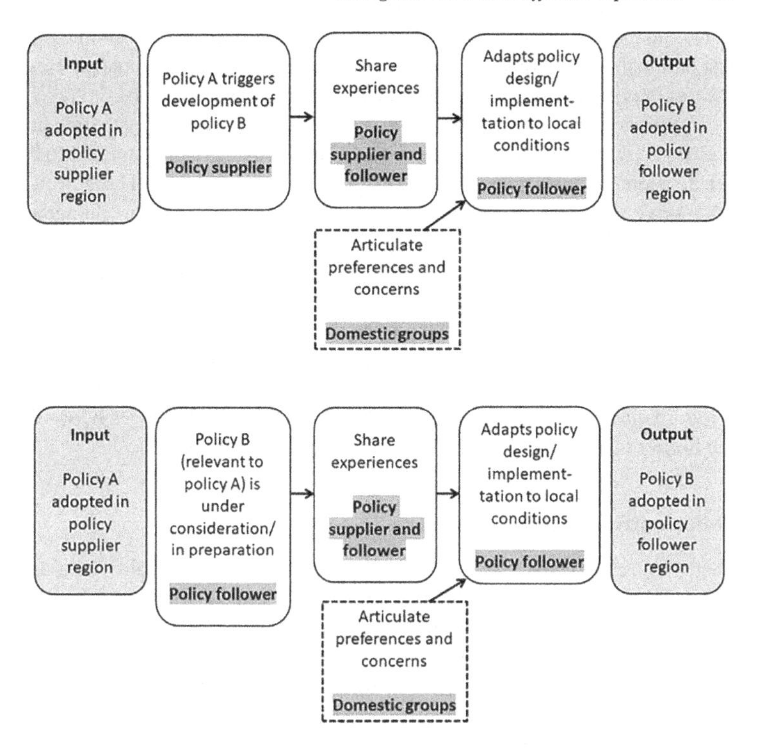

Figure 3.4 Two models of indirect diffusion.

Source: Author

Note: These charts use the major steps from learning diffusion for simplification purposes. Other diffusion mechanisms may also theoretically occur in the context of indirect diffusion.

development and facilitation. Led by the deputy secretary of the municipal government, it contains 15 institutional members (SEEE, 2016). Shanghai has issued a series of carbon finance documents, such as trading service provider management measures, CCER pledge loan business rules, and carbon margin trading business rules. Instead of treating carbon financial innovation as individual practices, Shanghai has taken a more sustainable approach, focusing on "supervision, risk management, and market cultivation. As such, the role model effect is high" (Interview, SH-E-030120).

Many regional governments in China are also actively pursuing green finance strategies. This has generated further opportunity for policy diffusion. As the main institution working on Shanghai's carbon finance, the SEEE has also helped other regions research and explore their policy options. These include Zhejiang, one of the national green finance pilot zones launched in

2017, and Dalian, the capital city of Liaoning province, which was among the first national low-carbon province and city pilots. In this way, national piloting policies in other relevant fields often help catalyze indirect diffusion.

In addition, Shanghai is frequently invited to speak at green finance events and workshops on topics such as cooperation on green financial product development and green system research and construction. These occasions serve as yet another channel for policy exchange between Shanghai and other regions.

Such extensive regional policy exchange and diffusion are not only attributable to regional factors. Indeed, green finance is a policy that has since 2015 been promoted at length from the top, by the central government and led by the powerful People's Bank of China (People's Bank of China, 2018). This provides significant motivation for regional governments to develop their own green finance policies. Once again, this points to the interlinkages between the horizontal and vertical dimensions of policy interactions.

Policy entrepreneurs

Shanghai's experience has shown that policy entrepreneurs also exist in China. First coined by American political scientist John W. Kingdon, a "policy entrepreneur" is an individual who takes advantage of opportunities to influence policy outcomes to further their own interests (Kingdon, 1984).[9] Several important policy entrepreneurs, largely from inside government, have contributed to policy innovation and diffusion regarding Shanghai's ETS.

The high-ranking secretary general of the Shanghai DRC, Zhou Qiang, has been one such person for Shanghai's pilot. His interest in developing Shanghai's ETS and garnering political support for its diffusion stem from his conception of the innovative and feasible nature of the instrument, believing that market-based policy can drive carbon mitigation (Interview, SH-E-271119; Interview, SH-E-051219). He was also actively engaged with relevant topics, for example by joining study trips to both international and domestic institutions to learn about carbon finance (Interview, SH-E-030120). The division director of the RCEPD of the Shanghai DRC, Ni Qiaglong, has played a similar role, also demonstrating quick learning ability, professionalism, and strong management skills (Interview, SH-E-011420; Interview, SH-E-030220). At the working level, Ling Yun from the RCEPD has been instrumental, acting for the office act as a bridge between the government and the supporting institutions (Interview, SH-E-051219).

As a new and complex policy for Chinese policymakers, ETS requires technical and nuanced expertise. The competence of these policy entrepreneurs and the value they place on ETS have facilitated policy progress, helping to ensure higher-level political support and delivering sound policymaking (Interview, SH-E-011420).

Notes

1 GHG emissions are not currently categorized as environmental pollutants in China.
2 From official assignment in 2011 until operation, the pilots had about two years to complete policy design, data collection, market infrastructure development, communication, and capacity building for covered entities.
3 All pilots were assigned national ETS capacity building centers, but no direct financial resources are provided by the central government. Those with more resources (such as Shanghai) are thus more active in carrying out activities.
4 See for example Heggelund et al. (2019).
5 The same approach is applicable to the national registry led by Hubei.
6 A good example is benchmarking. Shanghai has nine different benchmarks for the power sector, while the national government used only three to test allocation.
7 The Guangdong pilot, for example, has displayed this mechanism in its ETS diffusion with non-pilot regions Guizhou and Heilongjiang provinces (Interview, GD-E-170120; GD-E-200120).
8 Formerly Shanghai's financial policy supervisor; since 2018, the municipal Financial Supervision and Administration Bureau has taken this role.
9 Some Western researchers define policy entrepreneurs as located only outside the state, who "from outside . . . introduce, translate, and help implement new ideas into public practice" (Roberts & King, 1991, p. 147).

References

Chen, N. (2019, September). 区域温室气体控制目标与区域碳强度核算 [Regional GHG emissions control targets and regional carbon intensity accounting]. PPT presented at the capacity building activity for local authorities in Heilongjiang Province under the EU-China ETS Project. https://eu-chinaets.org/upload/file/20190919/1568880091667647.pdf

Chongqing DRC. (2018, January 4). 全国碳排放权交易体系正式启动重庆成为西部地区唯一参与全国市 场联建省市 [The national ETS officially launched and Chongqing became the only western province to participate in the joint construction of the national market]. Tanjiaoyi. http://www.tanjiaoyi.com/article-23591-1.html

Environmental Defense Fund & Energy Research Institute (ERI). (2018, May 23). *The Progress of China's Carbon Market 2017*. www.edf.org/climate/report-evaluates-chinas-national-carbon-market

Goron, C., & Cassisa, C. (2017). Regulatory institutions and market-based climate policy in China. *Global Environmental Politics*, 17(1), 9–120.

Heggelund, G., Stensdal, I., Duan, M. S., & Wettestad, J. (2019). China's development of ETS as a GHG mitigating policy tool: A case of policy diffusion or domestic drivers? *Review of Policy Research*, 36(2), 168–194.

Kingdon, J. W. (1984). *Agendas, Alternatives and Public Policies*. Boston: Little, Brown and Company.

Li, M. Y., & Hu, X. X. 李梅影 胡欣欣 (2012, August 21). 上海碳排放交易试点启动 宝钢等200家企业获免费配额 [Shanghai carbon emissions trading pilot program launched, 200 companies including Baosteel received free allowances]. *21st Century Business Herald*. http://finance.ifeng.com/news/corporate/20120821/6926580.shtml

Lo, A., & Chen, K. (2019). Policy selection of knowledge: The changing network of experts in the development of an emission trading scheme. *Geoforum*, 106, 1–12.

National Development and Reform Commission (NDRC). (2014, August 6). 国家发展改革委关于印发《单位国内生产总值二氧化碳排放降低目标责任考核评估办法》的通知 [Notice of the NDRC on Printing and Distributing the "Measures for Responsibility Assessment and Appraisal of Targets for Reduction of CO2 Emissions Per Unit of GDP"]. www.ccchina.org.cn/nDetail.aspx?newsId=47834&TId=60

People's Bank of China. (2018). 《中国绿色金融发展报告（2018）》摘要 [Summary of "China Green Finance Development Report (2018)"]. www.gov.cn/xinwen/2019-11/20/5453843/files/b61d608674b04494b3ae1aef76dd7b13.pdf

Roberts, N. C., & King, P. J. (1991). Policy entrepreneurs: Their activity structure and function in the policy process. *Journal of Public Administration Research and Theory*, 1(2), 147–175.

Shanghai Bureau of Ecology and Environment (Shanghai BEE). (2019). 上海市2019年碳排放配额分配方案（征求意见稿) [Shanghai 2019 Carbon Emission Allowance Allocation Plan (Draft for Comments)].

Shanghai Development and Reform Commission (Shanghai DRC). (2016, December). 上海市碳排放交易试点工作相关文件汇编. [Compilation of the Relevant Documents for the Shanghai ETS Pilot Work]. http://app.fgw.sh.gov.cn/images/jw_admin-upload-myupload_2567.pdf

Shanghai DRC. (2018). 上海市节能减排（应对气候变化）专项资金管理办法 2018 [Measures of Shanghai Municipality on the Administration of Special Funds for Energy Conservation and Emission Reduction (Climate Chang)]. www.shanghai.gov.cn/Attach/Attaches/201802/201802111001375244.pdf

Shanghai Environment and Energy Exchange (SEEE). (2014, July 30). 碳排放交易区域合作研讨会在沪成功召开 [Carbon emissions trading regional cooperation seminar successfully held in Shanghai]. *SEEE WeChat Account*.

SEEE. (2015, January). 上海碳市场报告（2013–2014) [Shanghai Carbon Market Report (2013–2014)]. www.cneeex.com/upload/resources/file/2018/07/16/25279.pdf

SEEE. (2016, February). 上海碳市场报告 (2015) [Shanghai Carbon Market Report (2015)]. www.cneeex.com/upload/resources/file/2018/07/16/25280.pdf

SEEE. (2017, April). 上海碳市场报告 (2016) [Shanghai Carbon Market Report (2016)]. www.cneeex.com/upload/resources/file/2018/07/16/25281.pdf

SEEE. (2018, April). 上海碳市场报告 (2017) [Shanghai Carbon Market Report (2017)]. www.cneeex.com/upload/resources/file/2018/07/16/25282.pdf

SEEE. (2019, April). 上海碳市场报告 (2018) [Shanghai Carbon Market Report (2018)]. www.cneeex.com/upload/resources/file/2019/07/02/26101.pdf

Shanghai Information Office & Shanghai Statistics Bureau. (2020). *Shanghai Basic Facts 2020*. http://en.shio.gov.cn/img/2020-ShanghaiBasicFacts.pdf

Shanghai Municipal Government. (2014, May 13). 上海市人民政府关于本市开展碳排放交易试点工作的实施意见 [Implementation Opinions of the Shanghai Municipal People's Government on the Pilot Work of Carbon Emission Trading in Shanghai]. www.cneeex.com/c/2014-05-13/487439.shtml

Shanghai Municipal Government. (2017a, March 1). 市政府关于印发《上海市节能和应对气候变化"十三五"规划》的通知 [Notice of the Municipal Government on Issuing the 13th Five-Year Plan for Energy Conservation and Climate Change in Shanghai]. www.shanghai.gov.cn/nw2/nw2314/nw39309/nw39385/nw40603/u26aw51762.html

Shanghai Municipal Government. (2017b, March 15). 上海市人民政府关于印发《上海市能源发展"十三五"规划》的通知 [Notice of the Shanghai

Municipal Government on Issuing the 13th Five-Year Plan for Energy Development in Shanghai]. http://news.bjx.com.cn/html/20170412/819711.shtml
Shanghai Municipal Government. (2019, June 17). 上海市人民政府办公厅关于成立上海市应对气候变化及节能减排工作领导小组的通知 [Notice of the General Office of the Shanghai Municipal People's Government on the Establishment of the Leading Group on Climate Change, Energy Conservation and Emission Reduction]. www.shanghai.gov.cn/nw2/nw2314/nw2319/nw10800/nw39221/nw44779/u26aw59347.html
Shanghai Statistics Bureau, (2019a, June 17). 2018年上海市国民经济和社会发展统计公报 [Statistical Communiqué on Shanghai's Economic and Social Development 2018]. www.shanghai.gov.cn/nw2/nw2314/nw2319/nw18462/nw44007/u21aw1388491.html
Shanghai Statistics Bureau. (2019b). *Shanghai Statistical Yearbook*. http://tjj.sh.gov.cn/tjnj/zgsh/nj2011.html
State Council. (2017). 国务院关于上海市城市总体规划的批复 [Reply of the State Council on Shanghai's Master Plan 2017–2035]. www.gov.cn/zhengce/content/2017-12/25/content_5250134.htm
State Grid Shanghai Electric Power Company. (2019). 上海市能源互联网白皮书. http://shoudian.bjx.com.cn/html/20190313/968514.shtml
Wang, E. D. (2014, August 19). 发改委正式启动碳强度考核，纳入干部考核体系 [NDRC officially launches carbon intensity evaluation and incorporates into cadre evaluation system]. *21st Century Business Herald.*, http://m.21jingji.com/article/20140819/036eb8e250cca892865b768333c73709.html
Zhang, M. S., & Wang, Y. 张敏思 王颖. (2018). 上海碳排放权交易试点政策体系和管理模式分析 [Analysis of the policy system and management model of the Shanghai carbon emissions trading pilot]. *Zhongugo jiangmao daokan (lilunban)*, 11. www.scicat.cn/finance/20190712/2667565.html
Zheng, S. et al. 郑爽等. (2014). 全国七省市碳交易试点调查与研究 [Investigation and Research on the Seven ETS Pilots in China]. Beijing: Zhongguo jingji chubanshe.

4 Hubei's ETS and diffusion experience

Shanghai's success in policy diffusion benefits from its abundant financial resources, high levels of transparency and openness, and strong governance capacity. Located in the less-developed central region of China, Hubei does not at first glance seem as equipped. Why, then, has Hubei also seen relative success in policy diffusion? Which mechanisms are dominant there? Which institutions or individuals have played an important role? This chapter unpacks Hubei's experience in an attempt to answer these questions.

Part of Central China, Hubei province sits at the junction of the Yangtze River Economic Belt from east to west and the Beijing-Guangzhou Railway Economic Belt from north to south. With a population of almost 60 million in 2019, Hubei is the tenth most populous province in China. It is made up of 12 cities, one autonomous prefecture, and 64 counties (Hubei Provincial Government Website, 2020). Due to its location, Hubei is the largest transport hub in Central China.

Hubei's economy is still growing and industrializing at high speed, much more so than Shanghai. During the design phase of Hubei's pilot, the province experienced a growth rate of over 13% annual GDP in the years 2010 and 2011. This then slowed slightly, landing at 9.7% in 2014, the year of the ETS launch. Hubei's annual growth rate was higher than the national average during this time (Duan et al., 2018) (see Figure 4.1). However, its per capital GDP is only slightly higher than the national average and is ranked bottom among the seven ETS pilot regions (Hubei Provincial Government Website, 2020, Interview, HB-E-100120).

In 2006, the State Council approved the Plan for Facilitating the Rise of Central China, positioning this region to be the national base for agriculture, energy, raw materials, modern equipment manufacturing, and high-tech industry, as well as China's transport hub. President Xi Jinping urged Hubei to take the lead in the rise of Central China during his visit to the province in 2013. In 2015, the central government launched the Belt and Road Initiative, and in 2016 the Yangtze River Economic Belt Plan and the 13th FYP for the Promotion of the Rise of Central China were published. In the same year, the CCP approved Hubei as an inland free trade zone (ibid.). These

DOI: 10.4324/9781003325307-4

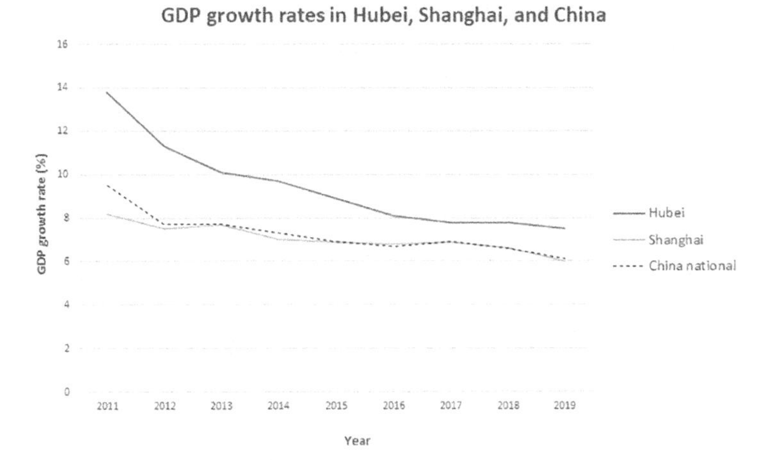

Figure 4.1 GDP growth rates in Hubei, Shanghai, and China.

Source: Author, based on Duan et al. (2018) and National Statistics Bureau website

initiatives provided continuous momentum for Hubei to develop its economy and industry.

Compared to the other pilot regions, Hubei has a unique industrial structure, made up of 8.3% primary industry, 41.7% secondary industry, and 50% tertiary industry in 2019 (Hubei Provincial Government Website, 2020). Its industrial production sector profile is made up of both large and small- and medium-sized industries, with a huge diversity of sectors, subsectors, and products (Interview, HB-E-260220). Its main industries include iron and steel, machinery, power generation, and automobiles. Hubei is also rich in agricultural, forestry, hydropower, and mineral resources. Its urbanization rate was 61% in 2019, slightly higher than the country's average, and Hubei continues to be the most urbanized province in Central China (Hubei Provincial Government Website, 2020). All this has led to significant and growing GHG emissions, to which industry is the largest contributing sector (Zheng et al., 2014, p. 211).

As in Shanghai, Hubei's emissions profile is not publicly available; thus, we use energy consumption as a proxy. Like the rest of China, Hubei's economy relies heavily on coal, which dominates its energy structure and accounted for 64% of its total energy consumption in 2010 (Hubei Provincial Government, 2017b). Together with oil (17%) and natural gas (2%), fossil fuels accounted for 82% of its energy mix in 2010, though its overreliance on coal has since lowered to 56% of energy consumption by the end of the 12th FYP period. During the same period, Hubei's energy intensity went down by 23%, and carbon emissions intensity fell by 20% (ibid.). Another 21% reduction of carbon emissions intensity was achieved during the first four years of the 13th FYP (Hubei Provincial Government Website, 2020). Although with a similar downwards trend as Shanghai, Hubei's energy and carbon intensities are at a much higher level, reflecting differences in the economic and industrial profiles between the two regions. Against this backdrop, Hubei is among the group of ETS pilots focusing on heavy industrial sectors in both their broader climate policy package and ETS.

ETS policy developments

Despite being the penultimate region to launch its ETS pilot, Hubei has also now completed a full policy cycle. However, unlike the Shanghai ETS, there are no distinctive trading periods in Hubei, reflecting the impression that pilots from China's central and western regions have weaker governance capacities than their eastern peers. Despite this, the Hubei pilot has witnessed a steep learning curve backed by strong political support, dedicated financial resources, and continuous learning-by-doing.

Like all the pilots, Hubei displays two-track ETS policy developments: alongside its own regional pilot track lies the national ETS track (see Figure 4.2). Like Shanghai, Hubei plays an especially important role in the national ETS. Not only is it one of the national carbon market training centers, but it also leads the development and operation of the national registry.

The Hubei ETS pilot shares a similar regulative system structure with Shanghai: a 1+1+N policy framework with one local government rule, one regulative document, and a set of supporting documents (see Table 4.1).

Both Shanghai and Hubei have similar regional government rules governing their offsetting mechanisms. Shanghai initially issued a three-year

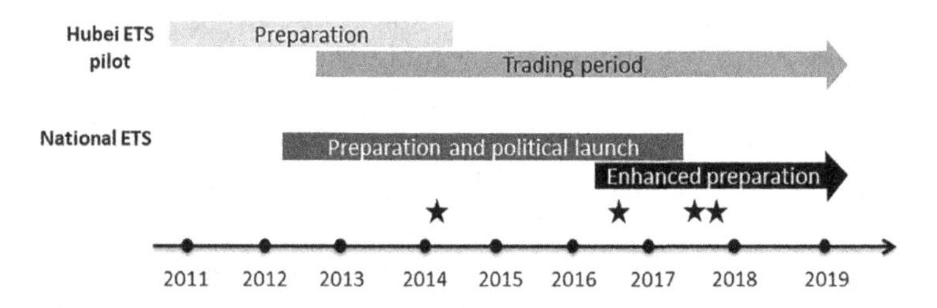

Figure 4.2 Policy development of Hubei ETS pilot.

Source: Author

Note: ★ indicate the following milestones:

Feb 2014: Hubei ETS pilot launched
Apr 2016: National ETS Capacity Building (Hubei) Center launched
Dec 2017: Political launch of national ETS
Dec 2017: Hubei selected to lead construction of national ETS registry

Table 4.1 Hubei's ETS pilot policy framework

Policy focus	*Document*	*Year published*
Overall	Interim measures for the management and trading of carbon emission rights in Hubei	2014, amended in 2016
	Implementation guidelines for Hubei ETS pilot	2013
MRV	Guidance on GHG Emissions monitoring, quantification and reporting in Hubei (trial)	2012
	11 sector-specific guidelines on GHG emissions accounting and reporting in Hubei (trial)	
	Hubei Province GHG emissions verification guidelines (trial)	2014
Allowance allocation	2014 Allowance allocation plan in Hubei ETS (including list of covered enterprises)	2014, updated annually
Market intervention	Management measures for injecting or buying back allowances from the carbon market	2015
	Interim measures for the administration of the collection and expenditure of the revenue from the carbon emission rights transfer	2015
Trading	CHEEX carbon emission trading rules (trial)	2014
	CHEEX carbon emission rights spot forward trading rules	2016
Offsets	Hubei DRC notice on the use of the offset mechanism in Hubei ETS	2015, updated annually

Source: Author, based on Duan et al. (2018)

allocation plan, later moving to annual issuance, while Hubei took an annual approach from the start. Both have general and sector-specific measuring and reporting rules and dedicated verification guidelines, but Shanghai has developed extra rules on allowance registration and management of third-party verifiers. In addition, Shanghai has a comprehensive package of trading rules while Hubei takes a lighter touch. Finally, Hubei has two dedicated rules regarding market stability, including the auctioning of allowances from a government reserve and injecting or buying back allowances. Such a focus on market management reveals a key characteristic of the Hubei ETS pilot: higher government intervention and more prominence of "the visible hand".[1] Shanghai, in contrast, takes a more market-oriented approach, letting the invisible hand of the market take the reins.

Hubei's institutional setup

Hubei's political dynamics underlie its selection for piloting by the central government, as one of just two regions (alongside Shenzhen) that proactively signaled interest in piloting ETS. Upon analysis of the state and trends of global ETS policy, in 2010 a local expert in Hubei proposed a vision to develop Wuhan into a carbon financial center for Central China or even for the whole country. The proposal was submitted to the provincial government and attracted the attention of the secretary of the provincial CCP committee. This then led to Hubei communicating its interest in ETS to the central government, before the latter had even begun considering testing the policy via pilots (Interview, HB-G-250220). When the NDRC began making such decisions in 2011, Hubei's earlier proactiveness, alongside its geographical representation and previous work relevant to climate policy, led to its selection as one of the pilots (Interview, HB-G-250220). There has been strong political support underpinning the Hubei ETS pilot from its initiation.

Hubei has outlined its specific ETS ambition in several strategic documents. Hubei's 13th FYP states that it "strives to build Wuhan into a Central China regional carbon trading and national carbon financial center" (Hubei DRC, 2016). The province's 13th FYP for Climate Change and Energy Conservation upgraded this and set a timeline to "strive to [achieve this] by 2020" (Hubei Provincial Government, 2017a). In specifying the date, Hubei's ambition became starker than that expressed by Shanghai's official document, which gave no indication of timeline. This political commitment to ETS is important to understanding Hubei's ETS policymaking and its role in ETS policy diffusion within China.

Hubei takes a looser institutional approach to governing its pilot than Shanghai, especially during the preparation phase (see Figure 4.3). Its institutional arrangement has changed over time, owing to changes in policy demand, the central government's influence, and institutions' choices. Like Shanghai, Hubei has seen high overall institutional and personnel stability.

Parts of Hubei's 1+1+N regulatory ETS framework were developed and endorsed by different actors within the provincial government and the main supporting institutions. The Hubei Provincial People's Government and its DRC share the role of ETS regulator.

Rather than establishing a new institutionalized management framework—an approach taken by Shanghai and Shenzhen—Hubei has drawn on its existing

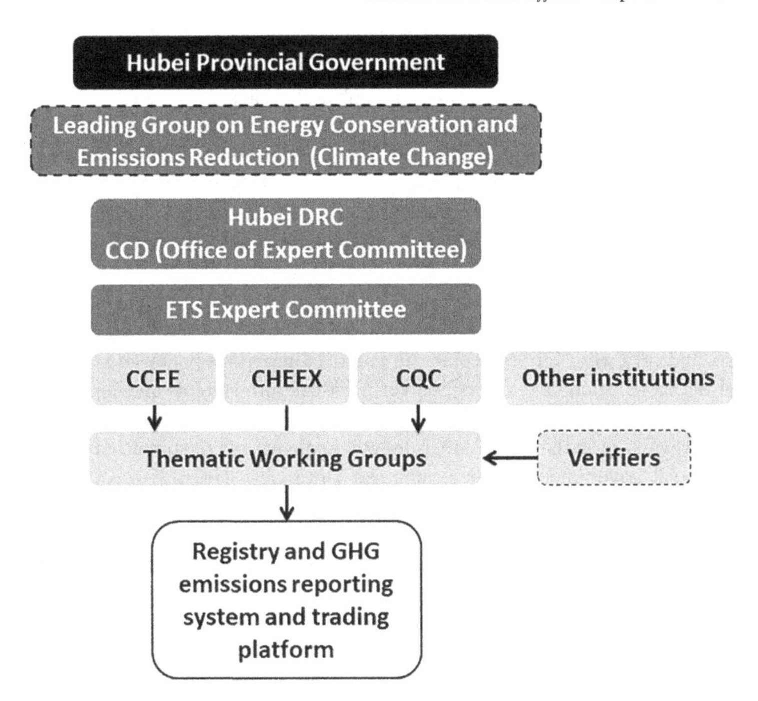

Figure 4.3 Institutional setup of Hubei ETS (in preparation and early operation phase).

Source: Author based on interviews with Hubei experts

Note: The larger dashed box indicates that this leading group provided high-level political guidance but was not directly involved in the ETS pilot. Although verifiers are listed as a separate box, the CQC was part of the third-party verifiers selected by the Hubei DRC from the beginning of the pilot.

climate change governance structure to administrate its pilot. In 2007, Hubei established the Leading Group on Energy Conservation and Emissions Reduction (Climate Change) led by the governor (Hubei Provincial Government, 2007). Climate change was in the parentheses of this group's title, meaning that it was under the institutional framework "one institution with two titles", a distinctive Chinese institutional feature.[2] It had two offices, one responsible for energy conservation and the other for (pollution) emissions reduction and climate change, respectively. The Hubei DRC hosted both and took charge of overall coordination and day-to-day operations. The then-Hubei Environment Protection Bureau helped oversee work related to environmental pollutant control. Although the Implementation Guideline of the Hubei pilot envisaged a new leading group dedicated to ETS, no public information could be found on its establishment. Instead, political guidance seems to have come from the existing leading group in charge of broader climate change issues.

This group, however, has much less direct impact on Hubei's ETS as compared to the newly established ETS leading group in Shanghai.

Furthermore, as in other pilots, the Hubei DRC acted as the pilot's competent authority. However, unlike some of its peers, including Shanghai, the Hubei DRC had a dedicated Climate Change Division (CCD), as opposed to subsuming it under a division with a broader set of responsibilities.

Another unique feature of Hubei was the significant role of the expert group in its ETS pilot development, established by the Hubei DRC in March 2016 (Hubei DRC, 2016). Its mandates included advising policymaking (particularly on ETS coverage, allowance allocation, and auctioning), monitoring market stability and providing relevant policy advice, evaluating offset use, and reviewing verified reports. In practice, important policy and technical decisions, such as choosing market infrastructure systems and formulating annual allocation plans, usually had to be approved by this committee. After review, proposals were either directly implemented or submitted to the provincial government for final approval (Interview, HB-E-270220; Interview, HB-G-250220). The office of this expert committee was located at the CCD of the Hubei DRC, while the China Hubei Carbon Emission Exchange (CHEEX) shared responsibility for its daily operation with the DRC (Hubei DRC, 2016). Prominent experts from regional academic, research, and market institutions and national technical institutions were part of the committee. External institutions also have been important for the Hubei ETS. In contrast, Shanghai relied primarily on its own regional institutions.

There were three main technical supporting institutions during the preparation and early operation phases of the Hubei ETS pilot: Wuhan University, the CHEEX, and the CQC.[3]

Around 2011, Wuhan University established the Climate Change and Energy Economics Study Center (CCEE) to research issues related to the low-carbon transformation. Headed by Qi Chaozhou, chairman of the ETS Expert Committee, the CCEE was essential in the preparation, establishment, and policy fine-tuning of the Hubei ETS pilot (Interview, HB-E-071219), especially in the earlier phase with its projects directly assigned by the Hubei DRC. National ministries also supported some of its projects. The CCEE was responsible for regulatory framework development, allowance allocation (including both pre-allocation and post-adjustment), compliance, and policy evaluation (Interview, HB-E-071219; Interview, HB-E-270220).

Like Shanghai's SEEE, the CHEEX was newly established as a dedicated exchange for environment and climate related trading, with political support from the regional government and business support from the long-standing regional property rights exchange. Approved by the Hubei Provincial Government, it was founded by the Wuhan Optics Valley United Property Rights Exchange (OVUPRE), in cooperation with three other shareholders: Wuhan Iron and Steel (Group) Company, Daye Nonferrous Metals Group Holding Co., Ltd., and Hubei Provincial Agricultural Means of Production Group

Holding Co., Ltd. Five new shareholders were added in 2017. As in the case of the SEEE, all of the CHEEX's shareholders are national or regional SOEs. The CHEEX's role in the early phase of the Hubei ETS was similar to that of the SEEE in Shanghai. It was heavily involved in the development of the ETS policy framework, allowance allocation, and trading rules. It was also in charge of carbon market operation and operated its market infrastructure. Unlike the SEEE, however, the CHEEX is more active in market infrastructure, operating all the three systems: the trading platform, the GHG reporting system, and the registry (Interview, HB-E-071219; Interview, HB-E-100120).

Unlike the two previous regional institutions for Hubei, the CQC is a national technical and service institution. As a quasi-government-affiliated institution, its carbon trading expertise stems from the CDM era beginning in 2007, and then further developed via China's domestic offsetting mechanism, the CCER, introduced in 2012. Based on these experiences, the CQC is familiar not only with MRV and the practices of these earlier market-based mechanisms, but also with industries' production and operation processes.

As a specialized third-party certification body, it also has experience in formulating verification rules, process documents, verification reports, and monitoring templates (Interview, HB-E-260220). Its ETS assignments were closely linked to its existing expertise. Like the SECERC in Shanghai, the CQC provides both policy advice for the government and third-party verification services for covered companies. It led the development of the pilot's MRV policy framework, including generic guidance on measuring and reporting, 11 sector-specific rules, and verification guidance. Other regulations and rules were led by other institutions, but the CQC provided feedback throughout. It also verified historical emissions data, which laid the foundation for defining the allowance allocation plan (Interview, HB-E-260220).

Other academic and research institutions, such as Huazhong University of Science and Technology, Hubei University of Economics, and the Macroeconomic Research Institute of the DRC, also contributed to the Hubei pilot. For example, Huazhong University forecasted Hubei's GHG emissions and energy consumption trajectories, necessary for setting the ETS cap (Interview, HB-E-270220). The Carbon Emissions Trading Collaborative Innovation Center of the Hubei University of Economics developed the sectoral reduction factor used in the annual allowance allocation process (ibid.).

Experts from the various technical supporting institutions worked closely together in sub-working groups on specific policy or technical issues. Among them, two were noteworthy. First, the Compliance Working Group, led by the CCEE and participated in by the CHEEX and other verification agencies (Interview, HB-E-071219). During the pilot's initial phase, pre-allocated allowances for the 12 covered sectors needed to be reviewed and adjusted. This extensive process resulted from the broad application of ex-post adjustment and price containment measures.[4] Taking the approach of "crossing the river by feeling the stones", rules related to this ex-post review and adjustment were vague and left much room for discretion. This group oversaw the ex-post adjustment process, all the way through to the settlement of company compliance (Interview, HB-E-270220).

The second was the Verification Technical Working Group, led by the CQC and participated in by other verification agencies such as the State Grid Electric Power Research Institute and the China Classification Society (Interview, HB-E-071219). The pilot's MRV rules were also drafted in highly generic terms. How they were implemented in practice required further clarification and coordination to ensure consistency. This second group was tasked with such responsibilities (ibid.).

Third-party verifiers help ensure environmental and market integrity. This role is to a large extent identical across different ETS pilots. However, while some have approved up to 20 verifiers, Hubei has only eight, citing the importance of ensuring data quality (Interview, HB-E-260220).[5]

Like in Shanghai, institutions and experts involved in Hubei's ETS policymaking have close *Guanxi* relationships with the government. Both Hubei and Shanghai faced complex tasks in designing and operating an ETS under severe time pressure. However, they chose different institutional arrangements to match their regional context, existing governance structure, and traditions (see Table 4.2).

Hubei's institutional framework over time

The institutional framework of Hubei's pilot has changed significantly since its implementation phase, driven by both internal and external factors. As in Shanghai, policy focus and demand have shifted, moving gradually from rule setting to carbon market operation and management. Consequently, Hubei's thematic working groups were dissolved after a few years of operation. Institutional choice also influenced Hubei's institutional evolution. More specifically, the CCEE continued to support the regional ETS until around 2018, but then shifted its focus from policy advisory back to academic research.

Table 4.2 Hubei and Shanghai institutional features

Institutional feature	*Hubei*	*Shanghai*
Dedicated leading group on ETS pilot	No	Yes
Centralized working relationship between government and supporting institutions	No	Yes
Reliance on ETS expert committee	Strong	Weaker
Involvement of nonlocal technical supporting institutions/ experts	Yes	No
Sub-working groups on ETS elements	Yes	No

Source: Author

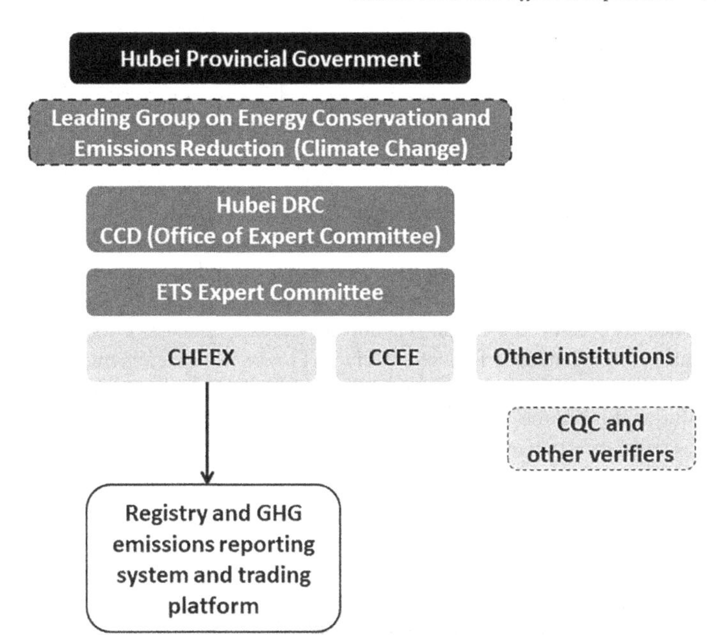

Figure 4.4 Institutional setup of Hubei ETS (mid-implementation phase).

Source: Author based on interviews with Hubei experts

Note: The increased size of the CHEEX reflects its stronger role in ETS operation. The exact role of the "other institutions" is not clear based on the obtained empirical data but is assumed to be stable.

Finally, central government restructuring was a significant external factor leading to a change in the governmental agency responsible for ETS at the regional level.

The major responsibilities of the three supporting institutions during the ETS pilot's operational phase are as follows (see Figure 4.4). Technical inputs, allowance allocation, and compliance are among the most important and politically sensitive elements of this phase. The CCEE supported and coordinated these interconnected processes until around 2018. The CHEEX, with a slightly stronger role than its counterpart in Shanghai, initially was central to supporting market infrastructure in Hubei, helping to manage the day-to-day operations of the ETS. It also supported the allowance allocation and compliance process (together with the CCEE).[6] The CQC, in contrast, moved away from policy advisory towards providing professional verification services to the covered enterprises.

After the CCEE shifted its focus back to academic research, the CHEEX took on supporting allowance allocation and compliance processes. Meanwhile, it has continued to manage the market side of the ETS. Like its Shanghai counterpart, the SEEE, it carries out a range of activities from rulemaking and -adjustment, operating market infrastructure, to developing market products and overseeing market risks. In addition, the CHEEX hosts the National ETS Capacity Building (Hubei) Center and is the body primarily responsible for the construction of the national registry.

The instructional framework of Hubei's pilot saw another major change in 2019 (see Figure 4.5). The restructuring of climate governance mandates was completed in Hubei in early 2019, whereby the CCD moved from the regional DRC to the Department of Ecology and Environment (DEE). In March 2019, there was subsequent adjustment related to the leading group, which was modified to become the Leading Group on Climate Change, Energy Conservation, and Emissions Reduction (Hubei Provincial Government, 2019), indicating a more prominent focus on climate change. The new scope of the institutional mandates meant that the responsibilities of the DRC and DEE in the daily operation of the leading group also changed. While the DRC was previously more dominating, a more equal division of labor between the two government bodies emerged.

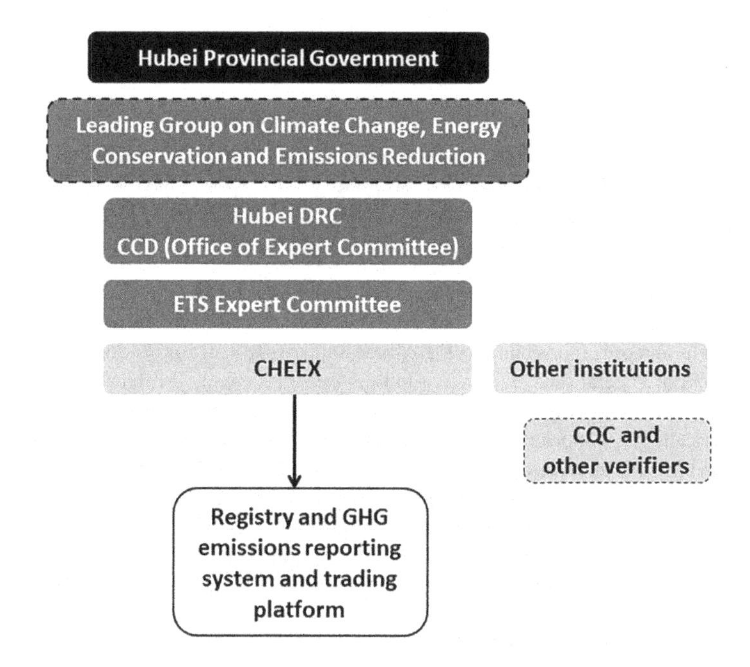

Figure 4.5 Institutional setup of Hubei ETS (later phase after the institutional reform).

Source: Author based on interviews with Hubei experts

Private sector lobbying

The private sector engages more prominently in lobbying in Hubei than in Shanghai with regard to ETS governance. The inclusion threshold of the Hubei ETS upon its launch was the highest (6–24 times higher) of the seven pilots. The size of companies covered under the Hubei ETS was thus often much larger. Of the 138 initially covered companies, over 40 were centrally owned SOEs and their subsidiaries, and approximately 20 were provincially owned (Qi & Wang, 2013). The large companies (i.e., central and provincial SOEs and other local enterprises) and industrial associations actively lobby the development of the Hubei ETS pilot. Experts from Hubei provide three examples from different sectors (Duan et al., 2018), outlined next.

Power is the largest sector covered by the Hubei ETS. During its first year of operation, the pilot employed benchmarking to allocate allowances to this sector. Benchmarking is often considered a much fairer allocation method in an ETS than grandparenting.[7] Though it entails higher demand on emissions and production data, the availability and relatively good quality of data alongside the homogeneity of products means the power sector is often a starting point for benchmarking.

However, the initial benchmark values used in Hubei led to more than 50% of power companies with a surplus of allowances, indicating the need for significant adjustment. A proposal was made at a meeting with private sector representatives: differentiate combined heat and power (CHP) units and tighten benchmark values. Several large power companies with CHP units not only opposed this during the meeting but also later submitted a written report, outlining their view of the proposed new benchmark for CHPs as too tight. This eventually led to the change of the benchmark value from the proposed level in their favor.

Iron and steel combined make up the second largest sector covered under the Hubei ETS. Large SOEs in this sector bargained with the government to include their different sub-companies in the ETS as one combined compliance entity rather than separate entities. This in principle goes against Hubei's ETS coverage rules, where companies with independent organization codes (i.e., independent legal identities) are viewed as a single compliance entity. Accordingly, sub-companies should be treated as separate entities. With this combined approach, companies wanted to exploit the "double 20" price containment mechanism and maximize compliance cost savings. Though the exact process leading to this decision is unknown, the government ultimately approved the proposal, granting them "special" treatment.

Due to the lack of quality data, grandparenting was used for the **cement** sector in the Hubei pilot during its first year. After acquiring verified emissions data throughout the first compliance circle, the government decided to move to benchmarking. Throughout the process of defining benchmarks, large cement companies raised their concerns regarding their energy and production data. The Hubei Building Materials Association also attempted to influence the technical supporting institutions' design of benchmarks. Though due in part to technical reasons such as the lack of a unified verification approach and changing emissions accounting methods, private sector influence here resulted in the predefined benchmarks for the compliance year 2015 being set too loosely. This led again to allowance surpluses across the sector. Further changes were made the following year: the predefined benchmark values were replaced by a predefined benchmarking level, i.e., the top *x*% of the sector's best performer. The benchmarks were now to be calculated at the end of the compliance year based on verified emissions data. Companies and sector associations once again tried to bargain with the government regarding the level at which this new approach should be set (Interview, HB-E-270220).

In this way, the private sector in Hubei has been seen to heavily influence the region's ETS policymaking process, for several reasons.

Firstly, Hubei is home to many large industrial companies. Its economy is characterized by a high growth rate, significant emissions, and industrial sectors accounting for more than 70% of its total emissions (Duan et al., 2018). This means that the emissions of individual companies in absolute terms as well as their share of covered emissions under the ETS are often large, resulting in enormous market power. Naturally, actors with such power are incentivized to influence allowance prices (e.g., via ETS design) to maximize their share in profits created by the ETS and/or to minimize compliance costs (Shim & Lee, 2016).

Secondly, central SOEs are directly under the authority of the central government and wield high administrative rank; many CEOs are given full ministerial rank, sometimes positioned higher than regional government officials (Eaton & Kostka, 2017). Central SOEs "dare to knock on the table" (Interview, HB-E-071219). Leaders of such companies are powerful, their influence extending high into the rafters of decision-making (Interview, HB-G-250220).

Thirdly, these SOEs contribute substantially to local tax revenue and employment. Their mark on the local economy is considerable. As such, their opinions have a significant impact on the regional government, which tends to protect them by covering up their negative environmental impacts.

Finally, the fierce lobbying and effectiveness of the private sector are also related to the governance capacities and traditions of Hubei. Unlike in the rest of China, eastern provincial governments are believed to have more effective administrative ability. Experts like to target Hubei for research on policy and governance issues to help garner insights on the national ETS. This is because of the similarity between Hubei and the national average development level as well as Hubei's overall government-enterprise relationship and the fact that the operation of government agencies reflects the national average. Hubei is less advanced than Guangdong or Shanghai, for example, and more advanced compared to other regions like those in North China (Interview, HB-E-270220).

Outside of lobbying, the Hubei government and their supporting institutions interact with private sector stakeholders through established channels. This is important considering the stark opposition that stemmed from the private sector during the early days of the Hubei pilot, when there was "lack of understanding" at the local, industry, and enterprise levels (Interview, HB-E-270220).

Unlike Shanghai's delicate and all-encompassing management approach, Hubei's interactions are less systematic and structured. For example, no dedicated personnel have been assigned to engage with specific sectors or companies. Investigative *Diaoyan*, requests for comment, and discussion symposia are the main channels through which the Hubei government collects feedback from companies and associations regarding policymaking and -improvement.

As MRV systems and allocation rules were being developed, the responsible technical supporting institutions visited selected companies and industrial associations. For example, the CQC conducted three studies when establishing the GHG reporting system (CQC, 2018b): (1) research and analysis of statistical data from the Bureau of Statistics; (2) visits to different government departments and industrial associations to analyze the emissions status of important energy-intensive enterprises; and (3) visits to enterprises by region and sector to better understand the management and quantification of emissions, their biggest challenges, and potential emissions reporting demand. Like in Shanghai, draft versions of the policy documents were sent to companies and associations for feedback (Interview, HB-E-260220). Feedback meetings also served as important venues to collect comments for decisions on allowance allocation (Interview, HB-E-270220).

Training has been used extensively to communicate policy. Improving stakeholder capacity by educating them about ETS and their roles and duties therein was essential especially during the preparation and early operational phase of the pilot. For example, the CQC conducted several training activities between March 2013 and November 2014, with a focus on MRV. Annual training sessions are also held, when government leadership provides direct communication and general ETS guidance and technical teams offer more specific coaching for companies and other stakeholders (Interview, HB-E-260220).

Existing administrative channels are also used for communication purposes. For example, after the annual allocation plan is discussed at the feedback meetings, the Hubei Economic and Information Committee as the government authority responsible for industrial sectors is consulted before the plan is finalized (Interview, HB-G-250220). Another example is that important policy documents are communicated through the vertical province-city-company channel, making use of the city's existing administrative capacities and the autonomous prefecture level governments (Interview, HB-E-260220). A third example relates to the management of SOEs. The national SASAC is authorized by the State Council to fulfill investor duties including oversight and evaluation of central SOEs. Similarly, the provincial

SASAC oversees provincial SOEs (Qi & Wang, 2013). In Hubei, the provincial SASAC and DRC jointly organized training activities for more than 3,000 company representatives between the start of the ETS pilot and April 2016 (CHEEX, 2016). Such joint sponsorship helped to secure private sector support for the ETS.

Verification agencies and the carbon exchange also maintain close contact with companies, acting as an informal and smooth communication channel between policymakers and the private sector (Interview, HB-E-260220).

Finally, companies across sectors in Hubei would call the technical team responsible for ex-post allowance adjustment in an attempt to influence the decision. This was in part due to Hubei's vague and not publicly available ETS rules, which left considerable space for interpretation. The lack of direct communication between the competent authority and companies on this sensitive matter also meant that the latter turned to the technical experts, hoping to use them as a bridge to send their messages to the government.

Institutional and personnel stability

Like in Shanghai, experts in Hubei have underscored the importance of high institutional and personnel stability as an important feature of the regional carbon market framework. For the technical institutions, this includes the director of the CCEE Qi Shaozhou, general manager of the CHEEX Liu Hanwu, and director of the CQC's Wuhan Branch Chen Weibin. All three institutions with their respective leadership have participated in the development of the pilot from the start (Interview, HB-E-260220).

The government side also sees a similar level of personnel stability. While in many regions (including Shanghai) government restructuring led to a change of competent authority ETS personnel,[8] Hubei saw the move of the whole climate change division from the regional DRC to its DEE, the team largely "moving together" (Interview, HB-E-260220).

Financial resources

Financial resources facilitate the development of new and innovative climate policies. Though less developed than Shanghai, Hubei also has a dedicated Special Fund for Low Carbon Economic Development for the 12th and 13th FYPs. The fund was set up in 2011 when Hubei already had special funds for energy conservation, the elimination of backward production capacity, and energy efficiency for buildings. The government added two items, namely low-carbon piloting and new energy development, with a total of RMB 170 million per year for all five items combined (Tan, 2011). The annual amount allocated to low-carbon activities is RMB 15 million per year since 2015, which supports ETS design and operation, capacity building, and other low-carbon pilot programs (Hubei DRC, 2014; Interview, HB-G-250220).

The provincial DRC and the Department of Finance jointly coordinate and supervise the operation of the fund. Local DRCs[9] select eligible projects within their jurisdictions to submit for further evaluation. Then, a review panel of sectoral and financial performance experts decides which projects receive financial support (Hubei DRC, 2014).

In absolute terms, the total amount of regional funding for ETS in Hubei is smaller than in Shanghai, as Hubei is also faced with the mismatch between workload and money (Interview, HB-E-270220). However, compared to regions without such dedicated funds, Hubei is still in a good financial position to advance its pilot. ETS funding has largely been channeled to research projects on carbon trading, market infrastructure, capacity building, and emissions verification (China Environment News, 2019).

Policy diffusion mechanisms

Like Shanghai, Hubei also sees a huge diversity of policy diffusion mechanisms across all three dimensions: horizontal, vertical, and indirect. There are, however, substantial differences between the two cases regarding the relationship between these mechanisms. Here, we unpack the individual policy diffusion mechanisms, focusing on several questions. How influential is each diffusion mechanism in Hubei? How have different types of diffusion occurred? Why and via which actors? How have these changed over time? We also delve into a comparison of the two cases, Hubei and Shanghai.

Horizontal diffusion

Learning is the most dominant diffusion mechanism in Hubei, as in Shanghai. It has occurred between Hubei and several pilot and non-pilot peers, horizontally and vertically. *Diaoyan*, capacity building, events, and information gathering have facilitated these processes, and both formal and informal channels are used. Also like the Shanghai case, the responsible government authorities and their technical supporting institutions drive diffusion.

General patterns of learning diffusion are quite different from Shanghai, however (see Table 4.3). Firstly, intra-pilot learning is dominated by one-way learning in the Hubei case while it is often a two-way process in Shanghai. Secondly, regarding learning between pilot and non-pilot regions, non-pilot regions are the main active parties in Shanghai, while both sides take the initiative in Hubei. Finally, there has been comprehensive alignment with national ETS rules in Hubei, which is not the case for Shanghai.

Mutual learning between Hubei and other pilots

Unlike Shanghai, which experienced a lengthy mutual learning process with other pilot peers, one-way learning dominates in Hubei; learning from other pilots has helped Hubei to improve its own ETS design and implementation.

Table 4.3 Comparison of learning diffusion in Hubei and Shanghai

Time period	*Learning type*	*Driver (Hubei)*	*Driver (Shanghai)*
2011–2014	Between pilots	Hubei	Both
2014–mid-2015	Between pilot and non-pilots	Non-pilots	Non-pilots
	Continuous between pilots	Hubei	Both
Late 2015/beginning 2016 onwards	Between pilot and non-pilots	Both	Mainly non-pilots
	Aligning with national rules	Hubei	n/a

Source: Author

Although all pilots began their preparations immediately after receiving the mandate from the NDRC in 2011, Hubei was the penultimate region to launch in April 2014, lagging about one year behind the others. This gave it a window of opportunity to learn from the earlier movers, who "explored the road ahead, [allowing] those behind them to avoid the same pitfalls" (Interview, HB-G-250220).

Market liquidity has been an important area to learn about for Hubei. In December 2013, the provincial Legal Affairs Office and the DRC jointly led a *Diaoyan* delegation to visit Guangdong and Shenzhen. Supporting institutions such as Wuhan University, the CHEEX, and the CQC also went along (CHEEX, 2013). One important lesson was that the transaction volume in Guangdong and Shenzhen was not very large at the beginning of the operation phase. Rules were subsequently introduced in Hubei to make its market transactions more active (Interview, HB-G-250220). Measures to enhance market liquidity are unique to Hubei. For example, it was one of the first pilots to allow individual (including foreign) investors to participate in the carbon market. It also ruled that only allowances with transactions could be banked, pushing companies with surplus allowances to trade.[10] Furthermore, its market participation and transaction fees are low (Duan et al., 2018).

Another point of learning has been auctioning. Shortly before its launch, the Hubei pilot arranged to auction its government reserve allowances, to activate the market and explore the carbon price. Allowance auctioning experiences from the EU and Guangdong were used to inform the auctioning design (Interview, HB-E-270220). However, this was a one-off instance of learning for Hubei, because though Guangdong continued to auction allowances, Hubei mostly relies on free allocation.

During the implementation phase, Hubei has continued learning about measures taken by other ETS pilots, particularly the more advanced systems in the East China region. For example, when designing power sector allocation in 2015, Hubei referred heavily to the benchmarking schemes of Shanghai and Guangdong. Following their example, Hubei set different benchmark values based on the size and type of generator units (Interview, HB-E-071219).

Hubei has mostly exchanged with the pilots in Guangdong, Shenzhen, Shanghai, and Beijing, though structural differences in their economies and industries in some cases limited the extent of learning. Other than Guangdong, they are megacities with a typical post-industrialized economic structure, unlike Hubei's industry-centric profile, constraining the applicability of their experiences to our case study region (Interview, HB-E-071219; Interview, HB-E-100120).

International ETSs have also been a source of policy learning for the Chinese pilots. Hubei has especially drawn on experience and lessons from the EU ETS, using it as a "blueprint" (Interview, HB-G-250220; Interview, HB-E-270220). This perspective, however, lies outside the scope of this book.

Diffusion between Hubei and non-pilot regions: preparing for the national ETS

Despite its late start, the Hubei ETS soon managed to catch up (Interview, HB-G-250220), a notion confirmed via interviews with experts from outside of the region. The Hubei ETS has several unique characteristics, such as its huge market size, strong regional leadership support, and high market liquidity (Interview, SC-E-180120). As in Shanghai, national level government officials have also acknowledged Hubei's achievements regarding its "groundbreaking" carbon market development and innovation (CHEEX, 2015).

Beginning about one year after its launch, the Hubei pilot has attracted various non-pilot regions to learn from its experience, thanks to its good market performance and stability (Interview, HB-E-071219). Another reason for Hubei to stand out as a region from which to learn ETS policy is the representativeness of its economic and emissions structure that led to a focus on industrial sectors in its ETS. Unlike Beijing, Shanghai, and Shenzhen, with developed economies and large shares of tertiary industry, Hubei's economy—and ETS—are industry-dominated. Most non-pilot provinces can benefit from this to understand how industrial enterprises fit within an ETS (Interview, HB-E-260220).

Shanghai has experienced two distinctive rounds of policy learning from two different groups of non-pilot regions, driven by various motivations. In the case of Hubei, however, non-pilot regions have been motivated largely by one thing: the status and prospect of the national ETS. Promoted by the central government as an essential policy tool to reduce emissions, carbon markets were highly politically visible during the 12th and (most of) the 13th FYP periods. Even before the political launch of the national ETS in late 2017, the expectation that a national system would be in place had already motivated many non-pilot regions to learn from existing ETS pilots. Such learning occurred in preparation for compliance with the national ETS once operational (Interview, HB-E-071219). Beginning 2016, the national government also arranged a series of tasks for the regional governments to deliver within the framework of national ETS preparation, which provided further incentives to learn (top-down, vertical diffusion). "Only looking at policy documents is far from being enough. Real life experiences from the pilots are thus of great value (for non-pilot regions)" (Interview, HB-E-260220). ETS is a complex and practice-oriented policy.

Like in Shanghai, *Diaoyan* is a dominant form of learning in Hubei. At least the following regions have conducted ETS *Diaoyan* with Hubei: Jilin, Shandong, Zhejiang and its Ningbo City, Guangxi, Jiangxi, Hebei, Inner Mongolia, Shanxi, Fujian, Xi'an City, Dalian City, Kunming City, and Guangyuan City. The main participants from the policy follower regions are the regional DRCs and technical supporting institutions, though sometimes regional exchanges also initiate *Diaoyan* without their DRCs. From the Hubei side, the regional government and technical supporting institutions mainly participate, with selected companies also involved in some cases to share their experience based on the demand and interest of the incoming delegations. Among Hubei's three main supporting institutions, the CHEEX plays the biggest role, hosting meetings and showcasing their trading platform.

Many topics are covered during these exchange sessions, including methodologies and cap-setting processes, allowance allocation, and compliance as a focus on the policy design side and how to achieve high levels of market liquidity as a topic of note on the market side. On the technical front, non-pilots are interested in how reporting, trading, and registry systems are set up and managed and how emissions are verified. Communication with the private sector and securing their support is another important topic (Interview, HB-G-250220; Interview, HB-E-270220; website news of CHEEX).

Some *Diaoyan* led to further study trips (e.g., Shanxi visited Hubei several times) and other forms of learning such as capacity building. For example, Hubei has responded to delegation requests and organized on-site training for regions such as Inner Mongolia province, Shandong province, Ningbo City, and Kunming City (ibid.).

Among non-pilot regions that have learned from Hubei, Fujian province stands out. Firstly, it is the only non-pilot region in China that has developed its own ETS. The motivation for policy learning is hence to understand how to design and operate such a comprehensive regional ETS, rather than to prepare for a national system like the other non-pilot regions. Secondly, Fujian has directly used some of Hubei's experience in its own ETS policy design. Such a traceable result of the policy diffusion process is missing for the regions that have exchanged with Hubei, making the traceable case of Fujian all the more valuable.

Fujian began preparations for its ETS in March 2016, launching in December that year (Interview, FJ-E-271219). The timeline to prepare the scheme was even tighter for Fujian than for the seven pilots assigned by the NDRC; it thus also conducted several exchanges with Beijing and Guangdong. Fujian delegations to Hubei were organized to inform its own ETS development and explore the potential for cooperation. The majority of *Diaoyan* took place during the preparation phase, and topics circled the ETS issues from policy design to system development and operation, to capacity building and private sector engagement.

Hubei has been the main source of learning for Fujian for several reasons (Interview, FJ-E-050220; Interview, HB-G-250220). Firstly, the combination of its market performance and efforts to promote its pilot as a success has led to Hubei's ETS having a good reputation in the eyes of non-pilot regional peers such as Fujian. Secondly, among the pilots considered to be more "successful", most have a municipality-based ETS, with different sectoral coverage as compared to Fujian. Hubei, on the other hand, shares many similarities regarding the type and size of covered entities. Thirdly, the Hubei DRC invited its Fujian counterpart to an ETS event early on, which laid the foundation for DRC-to-DRC communication. At the working level, the two carbon exchanges, the CHEEX and HXEE, have also kept in close contact. Fujian thus decided early on to use Hubei as a reference both for ETS market infrastructure and policy design. Though Hubei was used as a blueprint initially, Fujian's system has since changed substantially and drawn on the experiences of other pilots more systematically over time (Interview, FJ-E-050220).

Hubei has also made huge efforts to enhance its attractiveness and engage proactively with other regions, motivated by the prospect of developing into a national carbon trading center and the determination to win the competition to host national ETS market infrastructure. These high-level political goals set by the Hubei government changed its attitude towards external audiences, transforming Hubei from a passive player to being much more proactive in policy diffusion processes (Interview, HB-E-270220; Interview, HB-G-250220).

The most dominant form of learning that Hubei has used in this regard is capacity building. Like Shanghai, Hubei also received the mandate from the NDRC to set up a National ETS Capacity Building Center (the Hubei Center). Hosted by the CHEEX and with strong political backing and resources, the Hubei Center takes a holistic approach, establishing a 1+1+N framework to promote ETS capacity building activities. This framework consists of a physical training center (the Tan Hui Building),[11] an online training platform, and industry-specific training and demonstration bases, which together form the largest system of its kind in the country (Zhao, 2017). Since its founding in April 2014, the Hubei Center has organized many training sessions for companies and DRCs across China on topics such as the ETS compliance circle, corporate carbon strategies, offsets, and trading simulations. It has also organized sectoral training sessions, such as for iron and steel, building materials, petrochemicals, and chemicals. By 2019, more than 60 training sessions covering over 20 provinces and cities were carried out with more than 9,000 participants (ibid.). The achievements of the Hubei Center are comparable to its counterpart in Shanghai.

Hubei's other technical supporting institutions like Wuhan University and the CQC are also involved in the training activities. While the CHEEX organizes training and sources coaches, the other two institutions usually focus on providing experts to share their experiences and expertise accumulated via the Hubei pilot. Besides local trainers, national level expert resources have also been invited to cooperate with the relevant national institutions such as the NCSC, Tsinghua University, and SinoCarbon.

The Hubei Center, as in Shanghai, has also explored opportunities for institution-to-institution cooperation. It has signed Strategic Cooperation Agreements on ETS Capacity Building with China Nonferrous Metals Industry Association, China Petroleum and Chemical Industry Federation, China Electricity Council (CEC), China Building Materials Federation, Environmental Defense Fund (EDF), and CQC (EDF, 2016).

The Hubei pilot has also demonstrated an unusually high level of transparency. Despite its lack of a specialized information portal like in Shanghai, the CHEEX has compiled Hubei's ETS policies and documents and made them easily accessible online. The pilot also has a standardized and transparent market information disclosure mechanism. ETS policies, including administrative measures, allocation plans, and trading and innovative carbon business implementation rules; market transaction data including stock, increments, and decrements of market allowances; and information on intermediary institutions, are all publicly available (Futures Daily, 2018). This level of transparency not only benefits market players in Hubei but also helps broader audiences to become familiar with the ETS remotely. Public information is used often in combination with other forms of learning as a complementary channel of policy diffusion, like in the case of Shanghai.

Finally, the CCEE of Wuhan University has used academic opportunities to share Hubei's ETS experience, another channel of policy diffusion. For example, the CCEE participated in a large research project led by the China National Institute of Standardization on the "standards and systems to support ETS". Here, it shared the Hubei experience with other partner institutions, such as the Jiangxi Academy of Science's Energy Research Institute, Beijing University of Chemical Technology, and national construction and cement industry associations (Interview, HB-E-270220). Numerous academic conferences and seminars have also been used by the CCEE to disseminate the Hubei ETS experience (Interview, HB-E-071219). Compared to more formal, government-led channels, such informal settings can be more comprehensive and effective. These are content-focused exchanges that do not involve ulterior interests or power dynamics (Interview, HB-E-270220).

The **competition** mechanism is much more prominent in Hubei than in Shanghai, and even among the other pilots. Its ambitious and progressive attitude towards ETS, however, was not present from the outset, emerging instead around early 2016. By this point, Hubei had set a clear target, stemming from the national and regional levels, to become the national center for ETS and carbon finance.

From the national level, in 2015 President Xi Jinping put forward the New Development Philosophy, consisting of five development concepts: innovative development, coordinated development, green development, open development, and shared development (Xi, 2015). This for the first time put environmental protection and sustainable development at the top of the country's political agenda, as it served as the guiding principle for both the CCP and the state. In 2017, High Quality Development was put forward by the State Council in its Government Work Report to replace "fast growth" as a fundamental economic development target for the country (K. Q. Li, 2017).

The New Development Philosophy entered the national constitution through an amendment in 2018. During the earlier phase of the Hubei ETS pilot, the regional government was concerned with the potential negative impact of the scheme on its economy. The increasing significance of green and high-quality development on the national agenda provided a new narrative (Interview, HB-E-071219). This contributed to the change in the regional leadership's attitude towards ETS. Such top-down influence on the regional government's behaviors indicates once again the interlinkage between vertical and horizontal diffusion.

At the regional level, Hubei's ambition and competitiveness is the result of a deliberate internal advocacy, where the provincial DRC managed to raise the topic of ETS to the top of the political agenda and secure support from leadership (Interview, HB-G-250220). The following value propositions were used to "sell" the idea of ETS idea to regional political elites. Firstly, ETS is a powerful policy with real, cross-sectoral economic and societal impact. It involves a huge number of actors including companies, verification agencies, research institutions, consultancies, investors, and offset project developers, making its ripple effects extend much further than other climate or environmental policies'. Secondly, existing stock exchanges and trading centers in other parts of China demonstrate the benefits of a similar trading center in Hubei. These institutions have driven the development of a range of industries and contributed to regional economic growth. What Hubei lacked was exactly such a financial trading platform, which could raise its economic and political standing. Finally, unlike many other market-based environmental policies, carbon trading has a global dimension, which extends the especially political impact of an ETS (ibid.). Together, these arguments used in the advocacy process show that the consideration behind Hubei's competition for the national registry lay beyond immediate policy interest and short-term wins. Rather, it was founded in strategic interests calculated to secure mid- to long-term benefits for its industry, economy, and economic and political influence.

The DRC consistently advocated for ETS to provincial leaders across administrative levels from the department all the way through to the secretary of the provincial party committee and the governor. These efforts led to an acknowledgement of the necessity and significance of building the national carbon market center. Intensive internal advocacy led to Hubei's "aggressiveness" and determination to win the competition to host the national registry. It illustrates again how specific policy characteristics can influence diffusion and that policy entrepreneurial behaviors certainly exist in China.

Hubei's determination emerged in many forms. Firstly, it engaged regularly with the NDRC in Beijing, at both the climate change department and upper levels. Secondly, Hubei sent the executive vice governor to the in-person defense held in Beijing in front of a selection panel. Only two candidate provinces sent representatives of such high rank (Interview, BJ-E-090220; Interview, HB-G-250220; Interview, HB-E-071219). Thirdly, the Hubei government wrote to the NDRC, pledging full support in terms of funding and personnel, venues, and local policies to manage the construction of the national registry, if it were to be selected (Hubei DRC, 2019).

Hubei's efforts paid off. Its application document received the highest score from the national ETS market infrastructure selection committee (Hubei DRC, 2019). As the NDRC wanted two different regions in charge of the two systems, Hubei chose to focus on leading the registry over the trading platform, a decision approved by the NDRC. The Hubei DRC cited a few reasons for this choice, including the more comprehensive functions of the registry, how it could help Hubei become the national carbon market and carbon finance center and develop green and low-carbon industries, and the attractiveness of relevant service agencies, talent, and green technologies (ibid.). This again reveals Hubei's strategic thinking. Its priorities lay not just in the construction of ETS market infrastructure, but also in the development of a whole low-carbon industry. Developing the registry transformed Hubei's development from a technical into a political task and a source of political performance (Interview, HB-E-071219).

Other forms of competition identified in Shanghai also exist in Hubei. For example, competition between pilots to benchmark against each other's ETS and competition between the pilots' carbon exchanges to expand their business activities to other regions of China. These are examples of both weak and virtuous competition. For pilots competing to launch their ETS first, the extent to which this was present in Hubei is difficult to judge given the available data. Several experts have suggested that the Hubei DRC's original plan was to also launch its pilot in 2013, but that the failed attempt to link with Guangdong and weak data foundations hindered this.

Hubei's experience of the competition diffusion mechanism also shows that policy advocacy characterized by entrepreneurial behavior within the Chinese governance system is as important as in the West, though the focus in China is to secure support from upper political leadership rather than build a horizontal coalition of powerful players.

Like in the Shanghai case, **socialization** is a more minor example of diffusion in Hubei, usually occurring in conjunction with other more dominant mechanisms. When paired with learning, it takes the form of capacity building activities, in which the norms and merits of ETS are explained and promoted.

Socialization is also linked with the competition mechanism in the Hubei case. Once the goal of building the province into the national carbon trading center was set, Hubei needed to enhance its influence within the ETS community across China. This led to considerable publicity and outreach efforts. Unlike Shanghai whose publicity endeavors focused on promoting the broader concept of ETS, Hubei was more explicit about ETS as an accomplishment, sharing its "ETS success story".

The main channels used by Hubei in this regard were conferences—such as the Yangtze River Forum, the China-US Climate Summit, the C40 Conference, and side events at the UN COPs—and media coverage (Interview, HB-E-100120; Interview, HB-G-250220). National media outlets such as the CCTV and the People's Daily covered the Hubei ETS extensively, portraying it as one of the leading pilots. People's Daily even printed a full-page report titled "How Hubei became a leader in carbon trading", highlighting for instance the secondary market transaction volume, transaction value, and other indicators as "number one in the country" (Fu, 2015).

Both the regional DRC and the CHEEX were critical, with other technical institutions like Wuhan University and the CQC supplementing.

Like in Shanghai, **coercion** in both its carrot and stick forms exist in Hubei. For the former, besides promoting regional cooperation with its neighboring provinces, Hubei has also explored the possibility to link its ETS with Guangdong. Hubei and Guangdong are the only pilots that have explored direct two-way linking.

National government-imposed performance evaluation systems significantly impact ETS policy diffusion. Empirical data from Hubei is in line with findings from Shanghai, in particular that the pressure created by this evaluation as an impetus for ETS policy diffusion is limited. The system lacks "teeth" as a whole, and ETS only accounts for a small portion of the total score.

The remainder of this section unpacks Hubei's experience of promoting cooperation with other provinces in Central China to develop a regional carbon market and of exploring bilateral linking with Guangdong.

Central China regional cooperation

Upon entering the implementation phase of its pilot, Hubei reached out to its Central China regional peers about cooperating on carbon trading. The region is composed of Hubei and five other provinces: Shanxi, Anhui, Jiangxi, Henan, and Hunan. During the pilot's launch ceremony in April 2014, the Hubei DRC signed a series of Framework Agreements for Cross-Regional

Cooperation and Exchange on Carbon Emissions Trading with the DRCs of the other five provinces (CHEEX, 2014). Like Shanghai, it aimed to expand the scale of its ETS by allowing companies from the partner provinces to participate in its market (Interview, HB-G-250220). Hubei's idea was to develop a regional carbon market, with itself seated at the center (Interview, HB-E-071219).

As a first step, the CCER offset mechanism was used to attract surrounding provinces to such a regional carbon market. In 2015 Hubei opened its market to external partner regions by accepting their CCER credits. A quantitative limit was set of up to 50,000 tonnes of such credits for each compliance company. Partner regions welcomed this initiative (Interview, HB-E-071219).

However, its impact was limited. Only one cement company from Anhui successfully sold its CCERs to the Hubei market; this was the only case to ever utilize this cross-regional carbon transaction policy. The policy was also short-lived. Due to the accumulation of Hubei's own CCER projects and the shift of the province's policy priorities towards accelerating domestic poverty alleviation, Hubei closed its market to externally generated CCERs after just one year, instead only accepting domestic credits (Interview, HB-E-071219; Interview, HB-G-250220). It became clear that realizing and sustaining such a connected regional carbon market was difficult (Interview, HB-E-071219). Different regional and political interests must be coordinated and incorporated, resulting in the division of interests and risk of gaming. The process becomes highly political and complex (various Interviews). Hubei's experience here aligns with the conceptualization framework of the coercion mechanism that foresees little lasting effect of this type of policy diffusion. Under the Chinese governance structure, unlike in Western countries, national level policy decisions—even the perception of these alone—can pose significant obstacles for cross-regional vertical coercion diffusion.

Also contributing to this complexity is the uncertainty related to the national ETS and the CCER (Interview, HB-E-071219). Like Shanghai, Hubei's experience demonstrates that diffusion mechanisms taking the form of regional cooperation face many challenges that require much effort and resources from all sides to surmount.

Another building block for regional cooperation is information sharing and training. The Hubei government organized several regional carbon markets seminars for central and western China provinces and was invited to Jiangxi, Guangxi, Anhui, and Zhejiang to conduct on-site ETS training (CHEEX, 2016). Such activities helped other provinces become familiar with ETS and the specifics of Hubei's pilot. Of course, though useful for cross-regional cooperation, seminars are far from the establishment of a regionally integrated carbon market. These information-based activities tend to be technical, and with regional cooperation a highly political task, they require much more in-depth political discussion.

Bilateral linking with the Guangdong ETS

Hubei was unique in its attempt to explore bilateral ETS linking with Guangdong. The Hubei DRC signed a framework agreement for cross-regional cooperation with the Guangdong DRC at the launch ceremony of the Hubei pilot (CHEEX, 2014). However, discussions between the two regions on potential linking date to the earlier preparation phase of the pilots. The initial idea was proposed by several national ETS experts (Interview, HB-E-071219). Linking carbon markets can bring economic and environmental benefits such as enhanced cost efficiency, higher market liquidity, reduced competitiveness risks, streamlined administrative processes, and enhanced climate ambition (ICAP, 2018).[12]

Despite initial interest in linking, Hubei and Guangdong soon realized there were significant differences between the two ETSs, including the sectoral coverage, cap-setting, allowance allocation, and trading rules. Such differences were a result of both sides choosing to design their ETSs based on their respective socioeconomic conditions. These differences created significant technical and political obstacles to linking.[13] Both the Hubei and Guangdong DRCs chose instead to focus their resources on advancing their respective ETSs during the operation phase. Linking thus fell from priority.

Generally, linking ETSs presents significant challenges, many of them political. These include concerns over distributional impact, the scale and direction of financial flows, the partial loss of control over one's own ETS, and higher exposure to external shocks (ICAP, 2018). Fundamentally, linking two systems requires coordinating and aligning mutual interests. Historically, successful linking has occurred between economically and politically integrated or interconnected regions (e.g., the expansion of the EU ETS to the European Economic Area and its link with the Swiss ETS) or trusted partners with a long tradition of regional cooperation (e.g., the Regional Greenhouse Gas Initiative and the California-Quebec link) (ibid.).

Hubei's failed attempt at bilateral linking and the short-lived nature of the regional cooperation align with this book's conceptual framework on policy diffusion: any coercion diffusion takes time to materialize and any injected policy impact peters out quickly. Discussions on linking stalled early on (Interview, HB-E-071219). Momentum on the topic has not picked up since.

Emulation does not exist in its pure form in either Hubei or Shanghai. While Shanghai is perceived by its subnational government peers and the national government as a leading region with advanced innovative capacities and strong governance, this is not the case for Hubei. Geographically located in the lesser-developed part of China, Hubei's economic and social developments have not yet reached such high levels.

However, this regional division is evolving rapidly; Hubei is catching up with its East China peers. The impetus for such regional rebalance comes from both the central government and Hubei itself. National strategies like the 2006 Rise of the Central China Region and the 2016 Yangtze River Economic Belt Plan, which encompasses 11 provinces and regions,[14] have offered the more advanced provinces in these regions like Hubei opportunities to accelerate their development and enhance governance capacity. Within the Central China region, Hubei is already in the lead, with the highest GDP per capita and rate of urbanization. In recent years, Hubei has continued to move forward by benchmarking itself against more advanced provinces. Following Shanghai, Guangdong, Tianjin, and Fujian, Hubei was also assigned by the central government as a national free trade zone. Since 2016, Hubei is seventh in China in terms of GDP (Hubei Government Website, 2020). It has also made strides in technological innovation between 2013 and 2019, home to 12 national high-tech zones and over 6,500 high-tech enterprises (State Council Information Office, 2019).

Hubei's experience shows that the progression of traditionally less-developed regions in China should be understood not only by their endorsement of certain policies but also by improvements in governance capacity.

Vertical diffusion

Like in Shanghai, bottom-up policy diffusion towards the national government was also considered a high priority for Hubei, with more political significance than horizontal diffusion. "We hope that some of Hubei's [ETS] practices can be used in the development of the national carbon market. Hubei's social and economic circumstances are similar to the whole country, leading to a high level of replicability at the national level" (Interview, HB-E-071219). Almost all interviewed Hubei experts emphasized the similarity of economic development and industrial structure between Hubei and China as a whole, citing Hubei's advantage in promoting its ETS experience to national decisionmakers.

Hubei's greatest contribution to national ETS development was its direct participation in the construction of its market infrastructure. This is like the experience of Shanghai. At the end of 2017, Hubei was selected by the NDRC to lead the development of the national registry (EDF & ERI, 2018). The registry provides carbon asset confirmation, registration, and settlement. The Hubei provincial government set up a leading group for this purpose, indicating strong political and institutional backing. After the shift of ETS responsibility from the provincial DRC to the DEE, there are signs that such strong political support remains. Tasking Hubei with the construction and maintenance of the registry not only demonstrated the national government's trust in the province, but it was also an opportunity for Hubei to prove its competence (Environment Protection Magazine, 2019). In an interview in 2019, the director of the Hubei DEE promised that Hubei would cooperate closely with other regions to build a registry that is "strictly supervised, safe and stable, boasts comprehensive functions and advanced technology, operates efficiently, and with strong scalability" (ibid.). Like Shanghai, Hubei has also further supported the national government in developing registry management and operation rules (Interview, N-P-241219; Interview, HB-E-250220).

Compared to Shanghai, Hubei was more deeply involved in national ETS policymaking, acting as a quasi-supporting institution. Led by the national power sector association, the CEC, and in cooperation with other partners including two SOEs, a third-party verification and consulting company, and an international institution, the CHEEX participated in the drafting of the Technical Guidelines for the Carbon Emission Trading of Power Generation Enterprises (CEC, 2018). The CEC and CHEEX also partnered to develop the carbon market operation test plan for the power sector (ibid.).

Besides participation in policymaking, the Hubei pilot informed the national ETS via other channels, such as *Diaoyan*. Supporting institutions responsible for policy or technical elements of the national ETS and sector associations visited Hubei to learn from their experience. These include Tsinghua University, the NCSC, and SinoCarbon (Interview, HB-E-270220; Interview, HB-E-260220). Hubei's allocation methodologies, market risk management, and low-carbon poverty alleviation schemes provided rich experience for the national ETS to draw upon (Interview, HB-E-100120). Hubei also sent their experts to the national ETS preparation working group set up by the NDRC in 2017. Other channels seen in the Shanghai case like requests for comment and nationally organized trainings have been also observed in Hubei but are complementary venues.

Both Hubei and Shanghai contributed significantly to the national ETS, a role that has also been acknowledged by central government officials as having "achieved remarkable outcomes". Zhao Yingmin, deputy minister of the MEE, in 2018 expressed his hope that Hubei would continue to disseminate its experiences and inform the national system (CHEEX, 2018b).

Top-down diffusion is driven by the national ETS and other country-level policies. Like in the Shanghai case, these have demonstrated far-reaching effects on Hubei and how it interacts with its peers.

National ETS policy

National ETS policy has directly impacted Hubei's ETS design more than Shanghai's. In the later operation phase of the pilot, as the path forward for the national carbon market became gradually clearer and more of its measures and guidelines were published, Hubei moved increasingly closer to its design (Interview, HB-E-100120). Hubei's system began to integrate and align with the national system so that domestic enterprises could better adapt to upcoming national rules (Interview, HB-G-250220). ETS coverage and MRV are the two main areas where such policy alignment occurred.

In January 2016, the central government issued the Notice on the Key Works in Preparation for the Launch of the National ETS, in doing so organizing nationwide emissions reporting from companies and collecting emissions data from 2013 to 2015 (ICAP, 2016). The notice listed eight sectors to be covered by the national ETS with an inclusion threshold of 10,000 tce annual energy consumption: petrochemicals, chemicals, building materials, iron and steel, non-ferrous metals, paper making, power generation, and aviation. Based on this, in 2016 Hubei adjusted its ETS coverage from 60,000 tce to 10,000 tce for the sectors that overlapped with those covered by the national ETS. In the following year, it adjusted the inclusion thresholds for the remaining industrial sectors under its ETS to the level set by the national government.

The notice also contained technical MRV guidelines, including a detailed template for reporting verified GHG emissions and a reference guideline for the third-party verification process. There were initially some differences between these national guidelines and Hubei's local MRV rules; Hubei thus aligned its MRV methodologies accordingly. It also conducted an additional round of historical data verification based on the updated methodologies (Interview, HB-E-270220).[15] Some national experts doubt the extent to which this is a result of policy diffusion, instead citing that it may simply have been to avoid the parallel application of two different sets of standards. However, this policy alignment has not emerged in Shanghai; it is thus less likely that it stemmed only from such technical concerns.

In addition to direct impact on the regional ETS design, national ETS policy development also influenced how horizontal diffusion between Hubei and other regions functioned. This occurred in a similar way as in the case of Shanghai, composed of both positive and negative impulses.

The most prominent positive impulse that accelerated policy diffusion from Hubei to other regions was the assignment of Hubei as the national ETS capacity building center. As it was not considered a leading or advanced region, this was an important guarantee of credibility for Hubei when approaching its regional peers.

National policies also hindered regional policy diffusion. Hubei experienced a slowdown of policy exchange and diffusion since around 2018, due to governance restructuring and the subsequent brake on the pace of national ETS development (Interview, HB-G-250220). Interviewed experts confirmed this weakening of ETS momentum across different regions, although the level of impact varies. For example, Guangdong experts believe that their ETS is in fact in a good shape and that their market is highly credible; in addition, the promotion of green finance and other policy has provided new incentives for developing more carbon market products—all these have led to continued policy innovation and diffusion.

Despite such far-reaching top-down impacts, regional pilots have also shown some resistance towards the national ETS. In particular, they are committed to continuing to implement their own systems despite the introduction of the national market. The national ETS is expected to lead to the shrinkage of the market size of the seven regional pilots, with different scales of impact. As an industry-centered ETS, Hubei's system has significant overlap with the national ETS regarding sector coverage. The impact on its market size is thus significant. This is less of an issue for some other pilots like Shanghai with broad coverage of industrial and nonindustrial sectors. Regardless of what challenges may lie ahead, the seven regional carbon markets seem to all have an interest in continuing their operation. "This is about policy continuity and the credibility of the regional government. Furthermore, setting up and operating an ETS requires substantial investment from the regional government and the setup of dedicated systems and teams. All these should be preserved" (Interview, HB-G-250220).

National policies and arrangements beyond ETS

As introduced in the Shanghai chapter, there are three main ways that broader national policies and arrangements influence ETS policy diffusion at the regional level: top-level policy documents and directional guidance, experimentation and piloting, and the "counterpart help" mechanism (i.e., top-down pairing of provinces). Among these, the first and second are highly relevant for Hubei, illustrated by the following examples.

Yunnan is a southwestern Chinese province with significant renewable energy and forestry resources. It was selected by the national government to be one of the first low-carbon province pilots in 2010. Its capital, Kunming, was among the second group of nationally assigned low-carbon city pilots in 2012. The central government has also promoted the concept of "ecological civilization" since 2012. In response, Yunnan has developed a series of policy documents to support the development of low-carbon and ecological civilization, including the Outline of the Low Carbon Development Plan (2011–2020) and the Implementation Plan for Deepening the Reform of Ecological Civilization Systems. Yunnan has voluntarily committed to peaking its carbon emissions by 2025 and has advanced work on forestry offsets and MRV system construction.

In 2019, Yunnan was assigned by the national government as a free trade zone. The State Council outlined its tasks, including providing "ETS resource reserves".[16] This opened a new window of opportunity for ETS policy interaction with other regions. In July 2019, Hubei sent a delegation to Yunnan to exchange on several topics including ETS (Interview, YN-G-080120). The delegation was composed of representatives from the climate change division of the Hubei DEE, its Macroeconomic Research Institute, the CHEEX, and other supporting institutions, who discussed the policy formation of the Hubei ETS, homing in especially on innovative or region-specific elements (Interview, YN-G-080120). There has also been ETS cooperation and exchange beyond Hubei, such as between Yunnan and other ETS pilot regions. Shenzhen and Guangdong are two examples (ibid.). The Shenzhen DRC promoted cross-regional ETS cooperation with Yunnan in 2016, though due to the centralized development of the national ETS and differing regional interests, this did not yield concrete results. As for Guangdong, Yunnan has proactively learned from its ETS pilot experience, in particular about its MRV system and *Tanpuhui*, its provincial carbon offsetting mechanism (ibid.).

Another example is policy learning between Fujian and Hubei. Located on the southeast coast of mainland China, Fujian is the only non-pilot region that has developed its own ETS. Fujian used Hubei's ETS as a major reference for its own policy design and market infrastructure development. The origin of Fujian's efforts to develop its own ETS, which led to such cross-regional policy diffusion, was the national ecological civilization pilot policy, where Fujian was assigned as the country's first pilot to test this new concept. Leveraging this opportunity, the Fujian DRC obtained the NDRC's approval to develop its regional ETS. This mandate was anchored in the ecological civilization pilot implementation plan released in 2016 (Interview, FJ-E-050220). Another factor is that in the country's 13th FYP on Energy Conservation and Emissions Reduction released by the State Council, where "establishing and improving market-based mechanisms" was listed as a priority. This enhanced the appeal of ETS for the Fujian regional government, which regarded it as an important political achievement in 2016 (Interview, FJ-E-271219). Fujian's experience demonstrates again the strong interlinkage between vertical and horizontal diffusion mechanisms.

Fujian's ETS development process has led not only to intensive learning from Hubei, but also to other types of policy diffusion with other regions such as Guangdong. Thanks to its market performance and the similarity of economic structure and ETS coverage, the Guangdong ETS was of great interest for Fujian (Interview, FJ-E-050220). The Fujian-Guangdong learning process is similar to the Fujian-Hubei case. Other relevant policies such as its provincial offset program *Tanpuhui* are also part of the learning process. Fujian's positions in these processes have evolved over time. After launch, the Fujian ETS in turn became the target for other peers like Heilongjiang, Guangxi, and Sichuan to learn from. Many wanted to learn about its forestry-based offsetting program (ibid.). Later, Fujian was assigned as one of the country's ECQT pilots. This led to further policy interactions between Fujian and other ECQT pilots (Sichuan, Zhejiang, and Yunnan) and other regions like Jiangsu, Hebei, and Shandong (ibid.).

Fujian's two main technical supporting institutions, SinoCarbon and the Fujian Haixia Equity Exchange (HXEE), greatly influence these policy diffusion processes. SinoCarbon is a national consultancy with extensive ETS experience at the pilot and national level. It has facilitated Fujian's policy learning, making and tailoring policy proposals by summarizing and synthesizing carbon market policies from other regions and the national government (Interview, N-E-02072020). Fujian has used national rules on MRV and allocation as a guidance. Fujian has learned specifically from Beijing and Guangdong (ibid.) regarding other ETS design elements such as market stability mechanisms and offsets. This dynamic learning process indicates the coexistence of vertical and horizontal learning.

Fujian's second important technical supporting institution is the HEXX, which has also supported the Fujian government in its policymaking of several market-based mechanisms, such as pollution rights trading (PRT), ETS, and the ECQT. It also acts as the dedicated trading platform for all these mechanisms (Interview, FJ-E-050220). In this context, it has also played a unique role in the dynamic learning processes across these different policies.[17] Although different regional government agencies and departments in Fujian govern various policies, the overlap of the HEXX has enabled a continuous and often two-way learning process between Fujian and many of its peers. Besides formal government-to-government channels, regular communication and information sharing via informal channels have been critical, in particular those between the carbon or environment exchanges (ibid.), proving to be effective in enabling policy diffusion. Figure 4.6 provides an overview of such policy diffusion between Fujian and other regions.

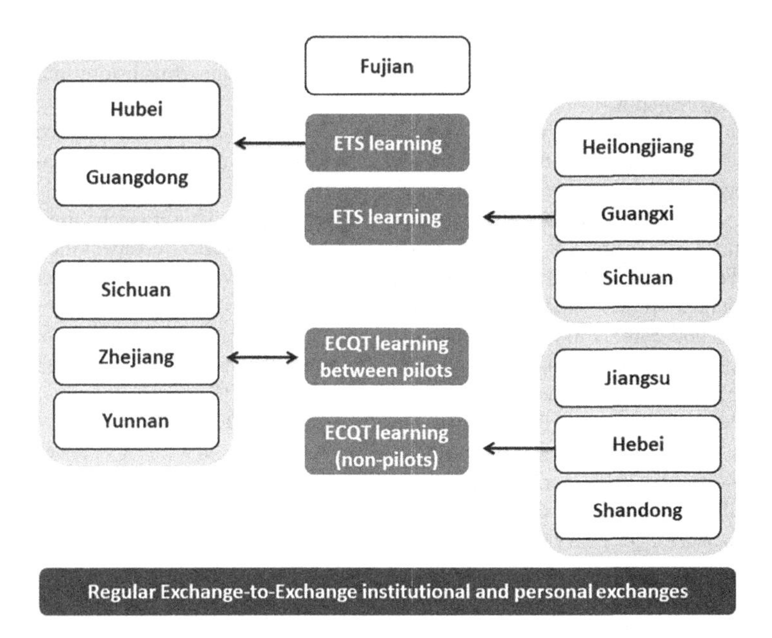

Figure 4.6 Policy diffusion between Fujian, Hubei, and beyond.

Source: Author based on interviews with Hubei and Fujian experts

Indirect diffusion

There are two main models of indirect diffusion. The first sees the development of policy B, as related to policy A in the policy supplier region. Policy B is then diffused to the policy follower region. In contrast, the second model sees the policy follower region using policy A from the supplier region as a source of learning for their development of policy B, which is similar or related to policy A.[18]

Taking ETS as policy A, interviews have revealed several related policies to be policies B. These include carbon and green finance, other market-based environmental or energy policies, *Tanpuhui*,[19] and forestry offsets. To a lesser extent, policies such as ecological compensation, agriculture offsets, and carbon neutrality are also relevant in different regions.

The Hubei government has promoted several policies related to its ETS. However, no substantial spillover effects into other regions (the first indirect diffusion model) have been able to be traced. For example, the Hubei government has developed several top-level policy documents that promote green finance, including the 13th FYP on Energy Conservation and Emissions Reduction and the Plan of Ten Strategic Measures for the Green Development of the Hubei Yangtze River Economic Belt (Hubei DRC, 2019; Interview, HB-E-100120). Hubei has also set the goal to build itself into the national carbon trading and carbon finance centers. Green and carbon finance policies have shown synergies with this ambition. Its technical supporting institutions also overlap with the ETS; for example, under the supervision of the provincial DRC and Financial Works Office, the CHEEX have not only pioneered carbon finance initiatives like the emissions pledge loans but have also supported the government in piloting the ECQT scheme.

Hubei has also promoted agriculture and forestry-based offset mechanisms in its poorer areas as a market-based approach to ecological compensation (Futures Daily, 2018). Overall, the Hubei ETS has served as an "important foundation and platform for . . . exchange and cooperation" regarding green and low-carbon industries, helping the province to explore new ideas in administrative systems and mechanisms (Interview, HB-E-100120). However, spillover effects of these policies have been constrained to within the province, not yet moving beyond its borders (Interview, HB-E-071219).

However, other regions have used the Hubei ETS pilot as a source of policy learning to develop other market-based policies: a case of the second indirect diffusion model. Hubei's neighboring province, Jiangxi, has been developing its own ECQT policy since 2018, looking to "relevant policy practices of other regions, among which we have mainly learned from the experience and lessons of the ETSs of Fujian, Guangdong, and Hubei" (Interview, JX-E-190220). Such ETS experiences have provided a reference for the development of elements of the ECQT policy, including its operating mechanisms, verification methods, quota allocation, transaction modes, and market adjustments (ibid.). This policy learning has come about, inter alia, via *Diaoyan*, online data collection, and telephone interviews (ibid.). A high-ranking delegation led by the deputy director of the Jiangxi DRC visited Hubei for an in-depth exchange in April 2018, joined also by the Ecological Civilization Division and Transportation Coordination Division of the DRC, the Industrial and Information Commission, the Environmental Protection Department, the Quality Supervision Bureau, and the Energy Conservation Center (CHEEX, 2018a).

Geographical proximity, the compatibility of their economic and industrial structures, and the general popularity of the Hubei ETS pilot contributed to the learning that occurred between the ETS and the ECQT. However, the existence of other regions like Fujian and Zhejiang that were assigned by the national government to specifically pilot the ECQT means it is likely that their ECQT experiences were a more direct source of learning as compared to Hubei's ETS. ETS pilots such as Hubei's have their own advantages in this learning process, however. For example, the length of the ETS pilots' operation leads to both a higher validity of their experience and huge diversity regarding the design of specific policy elements. Ultimately, limited empirical data makes it difficult to draw a full comparison between ECQT-ECQT direct diffusion and ETS-ECQT indirection diffusion.

As shown in both Shanghai and Hubei, indirect diffusion between regional governments could not have happened without the guidance from the national government. The central government's promotional efforts also incentivize regions to actively develop new policies like green finance and ECQTs, which then trigger cross-regional indirect policy diffusion. This illustrates again the vertical–horizontal diffusion interlinkage.

The indirect diffusion mechanism in the Hubei case is weaker than in Shanghai. This suggests that less-developed regions like Hubei require greater efforts to successfully promote policy diffusion than advanced regions like Shanghai. This must be backed by strong provincial leadership and substantial political, financial, and technical resources. This has been the case for direct ETS-ETS diffusion for Hubei, to which the region has attached significant political value. The lesser importance of indirect diffusion of ETS in Hubei is evident.

Policy entrepreneurs

Policy entrepreneurship appears more prominently in Hubei than in Shanghai. The necessity for political support and resources when less-developed regions promote new policies may explain why policy entrepreneurs have been so crucial in Hubei's ETS development and diffusion. These entrepreneurs exist within both the government and external institutions.

Zhen Jianqiao, deputy director of the Hubei DRC (directly supervising the CCD) and director of the provincial Energy Bureau, is a prominent ETS policy entrepreneur. His extensive experience in government, including his stint as a city mayor, has meant he is politically deft and courageous, exploiting the right tools and channels to achieve his objectives (Interview, HB-E-071219). In the early years of the ETS, he helped resist pressures from large companies and enforce compliance (Interview, HB-G-250220; Interview, HB-E-071219). He also convinced the executive vice governor of Hubei (also an ex-colleague) to travel to Beijing to represent the province during the selection process for the national registry, a move that proved critical in Hubei's victory (Interview, HB-E-071219).

Tian Qi, division chief of the CCD of the Hubei DRC as the lower-level leader, has also been significant, although with a less outgoing personality. At this level for many years, he is well versed in local politics, down-to-earth, and a quick learner (Interview, HB-E-071219; Interview, HB-G-250220). He is among only a few leaders responsible for ETS across the pilots in the role for such a long period: in his case almost ten years. Such long-term involvement, combined with his interest and ability, has allowed him to advance the ETS and Hubei's policy diffusion. Experts from Hubei have also pointed to similar characteristics and entrepreneurial behavior shown by several deputy division chiefs under Tian Qi.

Outside government, Liu Hanwu, general manager of the CHEEX, has been pivotal. Before joining the CHEEX, Liu worked at the Hubei Provincial Grain and Oil Group as a "provincially managed cadre".[20] He has introduced many innovative ideas on regional ETS (Interview, HB-G-250220). Several critical design features of Hubei's ETS are based directly on his proposals, such as the market adjustment factor, market stability measures, and trading rules (Interview, HB-E-071219). As a strategic visionary, he was an important advocate for Hubei to host the national registry (ibid.).

The characteristics of and behaviors shown by policy entrepreneurs in Hubei are similar to those observed in Shanghai. ETS requires specialized knowledge; the entrepreneurs' competence, expertise, and commitment to the instrument have allowed them to effectively implement the policy, advocate for it to top provincial leadership, and innovate policy.

Notes

1 The term "visible hand" was coined by Alfred Chandler (1977) and is used to describe how businesses can take control of their entire product chain (in contrast to Adam Smith's "invisible hand", i.e., free markets). This term has become popular in policymaking to describe government intervention.
2 According to the Office of the Central Organizing Committee, one institution means one legal representative, one financial account, one leadership, and one team; two titles entail two names, representing different identities when interacting externally.
3 Compared to Shanghai, there is little existing literature and public resources on Hubei's ETS governance. This chapter's insights thus largely stem from interviews

with experts who were heavily involved in the Hubei ETS across different policy stages.

4 One unique price containment measure is the "double 20" mechanism. If the difference between a company's pre-allocated allowances and annual emissions reaches either 20% of its pre-allocation level or 200 thousand tons in absolute terms, then its final allocated allowances are sealed at this level.

5 Other regions' considerations include helping build MRV capacity, resulting in higher numbers of verification agencies.

6 Exchange of staff between the two institutions also took place, e.g., a PhD student from the CCEE joined the CHEEX in 2017 and oversaw allowance allocation and compliance for that year (Interview, HB-E-071219). From then on, the CHEEX also took on more responsibility in the allowance allocation process (Interview, HB-G-250220).

7 Under grandparenting, covered entities receive allowances according to historical emissions in a base year/period. As such, grandparenting tends to reward historically high emitters. In contrast, under benchmarking, allowances are allocated according to performance indicators; efficient installations are thus rewarded.

8 At the regional level, the DRC is considered one of the most powerful government agencies. Officials are thus seldom willing to move to less powerful agencies such as the DEE or BEE. This led to many personnel across China previously responsible for climate change (including ETS) staying in the DRC even after the restructuring.

9 One level lower than the provincial DRC.

10 Many experts have criticized this measure, believing that mandatory trading as a condition for banking allowances goes against the rules of a free market—important for carbon markets.

11 "*Tan Hui*" translates literally to "carbon collection".

12 These are general benefits; the exact motivation for Hubei and Guangdong is not known. However, based on the economic and emissions profiles of the provinces, the flow of allowances would likely have been from Hubei to Guangdong (with an opposite financial flow) had the linking been successful. Such economic benefits may have been part of Hubei's motivation.

13 There are thus far only few successful cases of linking ETSs worldwide. In these instances, the design of their ETSs is either already aligned during the preparation phase in a coordinated manner or even jointly developed, or in the implementation phase by one system adjusting its rules to the other (see more in ICAP, 2018).

14 Three from East China (Shanghai, Zhejiang, and Jiangsu), four from Central China (Anhui, Jiangxi, Hubei, and Hunan), and four from West China (Chongqing, Sichuan, Yunnan, and Guizhou).

15 These 2013–2015 data had already been verified through previous compliance cycles.

16 The State Council published the China (Yunnan) Free Trade Pilot Zone Overall Plan in August 2019, which outlined key tasks for the province. However, the list of tasks is based on earlier submissions by Yunnan to the central government, meaning that the ETS element was less "imposed" on Yunnan, rather it stemmed from demand.

17 In the earlier piloting of PRT, the HEXX was also an important player regarding cross-regional policy diffusion, e.g., between Fujian and Zhejiang.

18 See graphical illustrations of these two models in the previous chapter.

19 A voluntary carbon credit mechanism focusing on non-ETS sectors, smaller enterprises, and/or individuals.

20 These cadres are appointed by the Provincial Communist Party Committee's Organization Department. For government officials, this title symbolizes political success.

References

Chandler, A. (1977). *The Visible Hand: The Managerial Revolution in American Business*. Cambridge, MA: Belknap Press, Harvard University Press.

CHEEX. (2014, April 8). 湖北碳排放权交易上线启动 [Hubei Launches ETS Pilot]. www.hbets.cn/index.php/index-view-aid-417.html

CHEEX. (2015, April 3). 国家发改委气候司李高副司长一行到我中心调研 [A Delegation Led By Deputy Director of the Climate Department of NDRC Li Gao Visits Center]. www.hbets.cn/index.php/index-view-aid-561.html

CHEEX. (2016, April 22). 湖北将建成全国碳交易能力建设培训中心 湖北模式由试点走向全国 [Hubei to Build National ETS Capacity Building Training Center; to go Nationwide]. www.hbets.cn/index.php/index-view-aid-636.html

CHEEX. (2018a, May 8). 江西省发改委刘兵副主任一行莅临湖北碳排放权交易中心调研 [Delegation Led By Deputy Director Liu Bing of Jiangxi Provincial DRC Visits CHEEX]. www.hbets.cn/index.php/index-view-aid-1330.html

CHEEX. (2018b, November 15). 生态环境部副部长赵英民一行莅临湖北碳排放权交易中心调研[Deputy Minister of MEE Visits CHEEX]. www.hbets.cn/index.php/index-view-aid-1456.html

China Electricity Council (CEC). (2018, November 11). 《发电企业碳排放权交易技术指南》编制说明 [Compilation Instructions "Technical Guide for Carbon Emission Trading of Power Generation Enterprises"]. www.tanpaifang.com/tanguwen/2019/0109/62800.html

China Environment News. (2019, February 13). 湖北：推进碳市场建设和碳排放交易[Hubei: Promote Carbon Emissions Trading and Carbon Market Construction]. www.cenews.com.cn/opinion/201902/t20190213_893253.html

China Hubei Carbon Emission Exchange (CHEEX). (2013, November 28). 省政府法制办、省发改委赴广州、深圳开展碳排放权交易立法调研 [The Legislative Affairs Office of the Provincial Government and the DRC Went to Guangzhou and Shenzhen to Conduct Legislation Research on Carbon Emissions Trading]. www.hbets.cn/index.php/index-view-aid-410.html

China Quality Certification Center (CQC). (2018b). 湖北省温室气体排放数据报送工作简介 [Brief Introduction of Reporting on GHGs in Hubei Province]. PowerPoint from the EU China ETS Cooperation Project Website. www.eu-chinaets.org/upload/file/20180828/1535453125284208.pdf

Duan, M. S., Wu, L. B., Qi, S. Z., & Hu, M. 段茂盛 吴力波 齐绍洲 胡敏 (Eds.). (2018). 中国碳市场发展报告—从试点走向全国 [Report on the Development of China's Carbon Market: From Pilots to National Market]. Beijing: Renmin chubanshe.

Eaton, S., & Kostka, G. (2017). Central protectionism in China: The "Central SOE Problem" in environmental governance. *The China Quarterly*, 231, 685–704.

EDF & Energy Research Institute (ERI). (2018, May 23). *The Progress of China's Carbon Market 2017*. www.edf.org/climate/report-evaluates-chinas-national-carbon-market

Environmental Defense Fund (EDF). (2016). 全国碳交易能力建设培训中心与美国环保协会签署合作协议 共同推进中国碳市场发展 [The National Carbon Trading Capacity Building Training Center and EDF Signed a Cooperation Agreement to Jointly Promote the Development of China's Carbon Market]. www.cet.net.cn/m/article.php?id=266

Environment Protection Magazine. (2019, April 8). 《环境保护》杂志专访全国人大代表、湖北省生态环境厅厅长吕文艳 (3) [Environment Protection Magazine's

Special Interview with Lu Wenyan, the Representative of the National People's Congress and Director of the Hubei Province Department of Ecology and Environment (3)]. https://m.huanbao-world.com/view.php?aid=95697&pageno=3

Fu, W. 付文. (2015, December 5). 碳交易，湖北如何当上领头羊 [How Hubei has become a leader in carbon trading]. *People's Daily*, Section 9. http://politics.people.com.cn/n/2015/1205/c1001-27892632.html

Futures Daily. (2018, December 16). 先行先试 "湖北经验" 领跑碳交易市场 [Try First: "Hubei Experience" Leads Carbon Trading Markets]. www.tanjiaoyi.com/article-25136-1.html

Hubei DRC. (2014, July 10). 湖北省2014年环保节能专项资金竞争性分配实施方案 [Implementation Plan for Competitive Distribution of Special Funds for Environmental Protection and Energy Conservation in Hubei Province in 2014]. www.21spv.com/news/show.php?itemid=5675

Hubei DRC. (2016, March 19). 省发改委办公室关于成立碳排放权交易专家委员会的通知 [Notice of the Provincial DRC Office on Establishing a Carbon Emission Trading Expert Committee]. www.tanpaifang.com/zhengcefagui/2016/031951534.html

Hubei DRC. (2019, June 26). 关于省政协十二届一次会议第20180604号提案的答复 [Reply on Proposal No. 20180604 of the First Session of the Twelfth Session of the Provincial Political Consultative Conference]. http://119.36.213.238:81/pub/fgw/fbjd/jytabl/201906/t20190626_747001.shtml

Hubei Provincial Government. (2007, September 11). 湖北省人民政府关于成立湖北省节能减排（应对气候变化）工作领导小组的通知 [Notice of Hubei Provincial People's Government on Establishing Hubei Province's Leading Group on Energy Conservation and Emission Reduction (Climate Change)]. www.hubei.gov.cn/govfile/ezf/201112/t20111208_1032136.shtml

Hubei Provincial Government. (2017a, July 3). 湖北省 "十三五" 节能减排综合工作方案 [Comprehensive Work Plan for Energy Saving and Emission Reduction during the 13th Five-Year Plan of Hubei Province]. www.hubei.gov.cn/govfile/ezf/201706/t20170615_1032939.shtml

Hubei Provincial Government. (2017b, October 28). 省人民政府关于印发湖北省能源发展 "十三五" 规划的通知 [Notice of the Provincial People's Government on Issuing the 13th Five-Year Plan for Energy Development in Hubei Province]. www.hubei.gov.cn/govfile/ezf/201711/t20171109_1221268.shtml

Hubei Provincial Government. (2019, March 20). 省人民政府办公厅关于湖北省节能减排（应对气候变化）工作领导小组更名并调整组成人员的通知 [Notice of the General Office of the Provincial People's Government on Change of the Name of Hubei Province's Leading Group on Energy Conservation and Emission Reduction (Climate Change) and Adjustment of its Composition]. www.hubei.gov.cn/govfile/ezbh/201904/t20190403_1387731.shtml

Hubei Provincial Government Website. (2020). *Information of Hubei*. www.hubei.gov.cn/

International Carbon Action Partnership (ICAP). (2016, January 25). *NDRC Outlines National ETS Sector Coverage*. https://icapcarbonaction.com/en/news-archive/338-ndrc-outlines-national-ets-sector-coverage

ICAP. (2018). *A Guide to Linking Emissions Trading System*. Berlin. https://icapcarbonaction.com/en/?option=com_attach&task=download&id=572

Li, K. Q. 李克强. (2017). 李克强在第十三届全国人大一次会议上的政府工作报告 [Li Keqiang's Delivery of the Government Work Report at the First Session of the

13th National People's Congress on Behalf of State Council] [Transcript]. www.scio.gov.cn/tt/34849/Document/1625877/1625877.htm

Qi, S. Z., & Wang, B. B. (2013). Fundamental issues and solutions in the design of China's ETS pilots: Allowance allocation, price mechanism and state-owned key enterprises. *Chinese Journal of Population Resources and Environment*, 11(1), 26–32.

Shim, S., & Lee, J. (2016). Covering indirect emissions mitigates market power in carbon markets: The case of South Korea. *Sustainability*, 8(6), 583.

State Council Information Office. (2019, June 19). 国新办举行"长江经济带建设与湖北高质量发展"发布会图文实录 [State Council Office holds press conference on "Construction of the Yangtze River Economic Belt and High-Quality Development in Hubei"] [Transcript]. http://www.scio.gov.cn/xwfbh/xwbfbh/wqfbh/39595/40712/wz40714/Document/1657141/1657141.htm

Tan, Q. L. (2011, August 13). 低碳试点的湖北行动 [Hubei's Actions for Low Carbon Piloting]. www.hb.chinanews.com/news/2011/0813/87696.html

Xi, J. P. 习近平. (2015, October 29). 在党的十八届五中全会第二次全体会议上的讲话 [Speech at the Second Plenary Session of the Fifth Plenary Session of the Eighteenth Central Committee of the Chinese Communist Party] [Transcript]. http://cpc.people.com.cn/n1/2016/0101/c64094-28002398.html

Zhao, M. 赵孟. (2017, February 9). 湖北探路碳交易：已成为全国最大碳市场，试点经验辐射多省 [Hubei explores carbon trading: It has become the nation's largest carbon market, and the pilot experience radiates many provinces]. *The Paper*. www.thepaper.cn/newsDetail_forward_1614668

Zheng, S. et al. 郑爽等. (2014). 全国七省市碳交易试点调查与研究 [Investigation and Research on the Seven ETS Pilots in China]. Beijing: Zhongguo jingji chubanshe.

5 Conclusion

The role of policy diffusion in China's climate transformation

Humans have become the driving force in our planetary ecosystem, now under existential threat from climate change. This calls for an all-encompassing transformation from our current societies and economies to an alternative carbon-neutral and sustainable development pathway. Transformation must happen across all geographies and political levels and be led by both the public and the private sectors alongside civil society. China has great transformation leadership potential. An economic and emissions heavyweight, it has already begun its transformation. The government is carefully studying and testing new concepts and instruments and working on many fronts to mitigate GHG emissions and transform its energy and economic systems. Whether China will be successful in this endeavor depends not only on good policy but on good governance.

This book has attempted to portray how and why low-carbon transformation policies spread across China's subnational regions and between governance levels. It has drawn on the policy diffusion framework, exploring how different diffusion mechanisms influence the ways that Chinese regions introduce climate policy. Its empirical case studies on ETS in Shanghai and Hubei enrich understanding of policy diffusion mechanisms based on innovative market-based climate policy in China. They also provide valuable insights into the country's overall governance system, often still a murky area of research both in the West and in China.

Shanghai and Hubei: a comparison of diffusion patterns

Like Shanghai, Hubei is relatively autonomous at the local level. However, there are significant differences between Hubei's and Shanghai's stage of development, industrial and energy structure, and governance traditions. Interestingly, Hubei has achieved a similar level of success with the diffusion of ETS. Like Shanghai, it has benefited from a highly stable institutional framework composed of committed government bodies, professional technical supporting institutions, and dedicated financial resources. A major difference, however, is that Hubei managed to secure political support for its ETS development and

DOI: 10.4324/9781003325307-5

Table 5.1 Comparison of diffusion mechanisms in Hubei and Shanghai

Diffusion mechanism	*Hubei*	*Shanghai*
Learning	Most dominant mechanism	
	Learning among pilots: Hubei takes more initiative; more one-way learning	Learning among pilots: initiative from both sides; mutual learning
	Learning among non-pilots: Hubei more active since 2015	Learning among non-pilots: non-pilots consistently the proactive driver
Competition	Stronger role (additional dominant mechanism)	Weak role
Socialization	Minor mechanism connected with learning (Hubei also with competition)	
Coercion	Top-down evaluation system with limited impact	
	Regional cooperation with limited success	
	Bilateral linking explored with little success	No bilateral linking efforts made
Emulation	Mechanism with least relevance; no 1:1 emulation	Shanghai is considered a leading region, though no 1:1 emulation
Bottom-up	Influences market infrastructure and related national ETS policy development	
Top-down	Alignment with national ETS rules	No alignment with national ETS rules
	National ETS policy and prospects impact horizontal diffusion	
	Comprehensive or policy-specific experimentation or piloting impacts horizontal diffusion	
Indirect	Model two of indirect diffusion more dominant	Model one of indirect diffusion more dominant

Source: Author

diffusion from top leadership. Shanghai, in comparison, is more outstanding in its governance capacity and degree of openness.

The Hubei case has also reflected that policy diffusion under the Chinese governance framework tends to happen both vertically and horizontally. Although with similarly diverse diffusion mechanisms, the relationships between those in Hubei and the patterns they display differ from Shanghai's (see Table 5.1).

Diffusion mechanisms

Following recent trends in the policy diffusion literature, this book treats the process as complex and multifaceted. Instead of setting a dichotomous, categorical dependent variable (adoption or non-adoption of policy), it recognizes the possible diffusion of policy *segments*, ranging from ends to means, policy inputs and outputs, to policy elements. In China, both horizontal and vertical diffusion are significant, given China's central-regional government structure and its long history as a unitary state.

Diverse forms of diffusion lead to the spread of innovative policies across China. There are strong links between vertical and horizontal dimensions of policy interaction, which entail a chain of drill-down effects. These are triggered by the central government's decisions and lead to policy adjustment at the regional level.

Horizontal diffusion

At the horizontal level, the experiences of Shanghai and Hubei confirm that different (though not all) types of policy diffusion have occurred across the ETS lifecycle. Differences in dominance and length of effect exist both within and between the cases of Shanghai and Hubei. Figures 5.1 and 5.2 provide an overview of these.[1]

The Shanghai and Hubei cases, alongside observations from experts and practitioners from various levels, have revealed the following insights regarding the different diffusion mechanisms in the context of ETS in China.

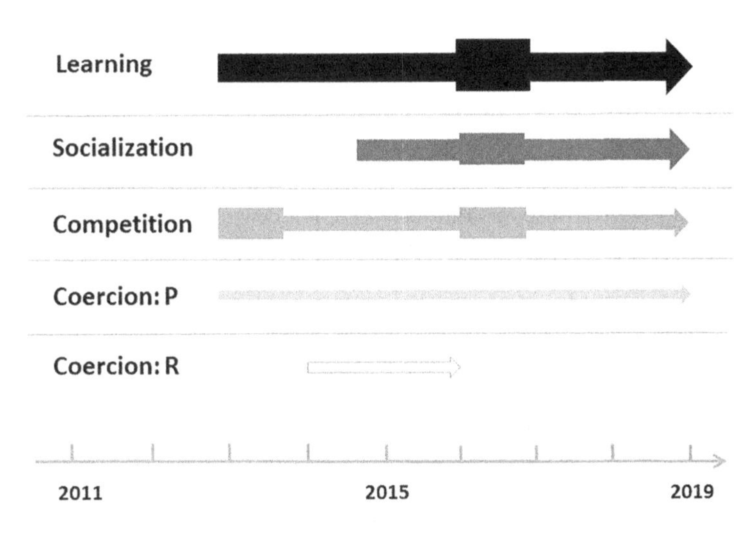

Figure 5.1 Horizontal diffusion mechanisms: Shanghai.

Source: Author

Note: P = performance evaluation system; R = regional cooperation. The order of mechanisms reflects their overall importance (most to least: top to bottom). The thickness of the arrows indicates the level of dominance in relative terms.

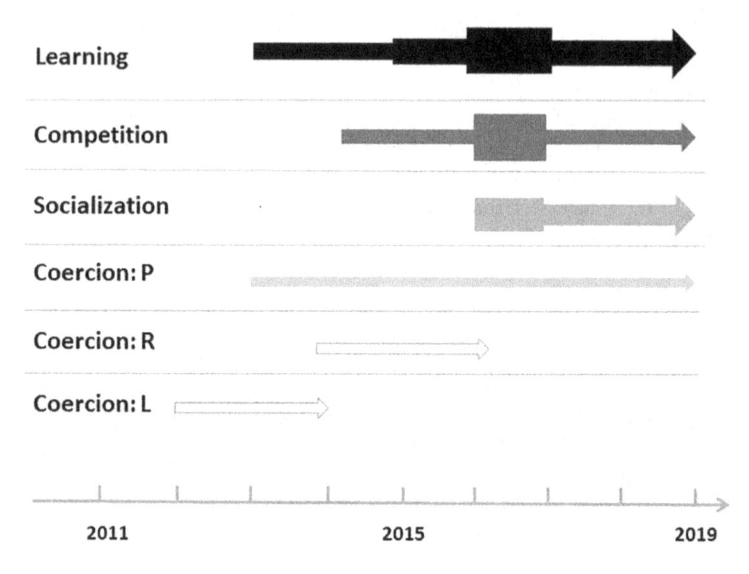

Figure 5.2 Horizontal diffusion mechanisms: Hubei.

Source: Author

Note: L = *linking*.

Learning is the dominant diffusion mechanism both for pilot-pilot and pilot-non-pilot regions. This takes different forms, including *Diaoyan*, capacity building, workshops and events, and various information channels. It occurs at several stages of policymaking. ETS is complex; learning thus covers a spectrum of topics both macro and micro, ranging from design specifics to operation measures, to the trading side of the carbon market. Both one- and two-way learning are possible, with the latter mainly between more advanced pilots, e.g., Shanghai and Guangdong. Learning from which region depends on who is considered "advanced" and whether regional conditions are compatible. This choice may also depend on specific policy elements, such as MRV (Beijing, Shanghai, and Guangdong), benchmarking (Shanghai and Guangdong), and auctioning (Guangdong). General policy design may also determine the choice of policy supplier region, e.g., a market-oriented versus government-dominated approach. The initiators of learning are usually the policy follower regions' governments, though carbon exchanges in policy supplier regions are sometimes the more active diffusion drivers due to their business interests.

Competition is another dominant mechanism that mainly occurs between pilots. The combination of piloting, the nomenklatura system, and the cadre evaluation system leads to virtuous competition, which surfaces predominantly as pilots benchmarking themselves against each other and consequently refining their ETSs. The level of competition varies across pilots—Hubei is significantly more competitive than Shanghai, driven by its ambitions to develop into a national carbon market and carbon finance center based on strategic economic and political considerations beyond ETS.

As a secondary mechanism, **socialization** usually takes place in conjunction with the dominant diffusion mechanism(s), learning and/or competition, primarily via capacity building or publicity activities. The initiators of this diffusion are mainly the policy supplier regions, i.e., the ETS pilots, focusing on either broader concepts of ETS or specific ETS pilot "success stories". For example, Shanghai hosts a biennial Green and Low Carbon Development Summit to promote ETS, and Hubei has gained extensive national media coverage on its success.

Coercion takes place in both carrot and stick form. A climate mitigation policy performance evaluation system organized by the central government creates pressure on regional governments to advance their energy conservation and climate change efforts (stick). However, carbon markets form only a small part of the evaluation matrix, and the system's accountability framework is relatively weak, limiting the coercion effect. Pilot regions also try to use (elements of) ETS to attract the cross-regional ETS co-development or even linking (carrot). Here, there is usually an asymmetry of power between the initiator and follower regions. Shanghai and Hubei have both promoted carbon market cooperation with their regional peers. Given China's diversity, provinces may be both proactive and passive regarding initiating policy diffusion, depending on their diffusion partners. Complex politics and heterogeneous interests alongside national ETS development mean success is rare and hard to sustain. This form of coercion tends to occur at the earlier stages of an ETS.

Emulation in its pure form does not exist. However, regions do look to "advanced" policy regions for ideas or specific measures that might help them meet their own policy objectives. Non-pilot regions usually look to the eastern pilot regions such as Shanghai, Guangdong, and Shenzhen. Whether the policy supplier regions share a similar economic and emissions profile with the region seeking policy ideas is another factor. Pilots with a more industry-based emissions structure such as Hubei and Guangdong are more attractive to most non-pilot regions still in the industrialization process, given the comparability of circumstances.

Geographical proximity contributes to policy diffusion, usually via policy networks that already exist thanks to economic or regional development strategies. Examples include the Yangtze River Delta region of which Shanghai is part, the five central provinces of which Hubei is part, the five southwest provinces, the Guangdong-Hong Kong-Macao Bay Area, and the Pan-Pearl River.

Good personal and institutional relationships between supporting institutions across regions also help policy diffusion, complementing formal channels of communication.

Vertical diffusion

Shanghai and Hubei have both attached more importance to bottom-up diffusion. Table 5.2 provides an overview of the different ways this bottom-up influence has been exerted and by whom. In most cases, this is directly organized by the national government with regional governments in a reactive position. Where the construction of market infrastructure and rulemaking was involved (perhaps also the most influential case), regional governments (supported by carbon exchanges) were more proactive, though the central government retained final decision-making power. Technical supporting institutions across national and regional levels have also in several instances played a more important role than the governments. Bottom-up diffusion follows the logic of learning, where regional expertise and experience inform central government decisions on the design or operation of policy elements.

Vertical top-down diffusion has also occurred in the cases of Shanghai, Hubei, and elsewhere in China, underlining the vast influence of the central government on policy behaviors and interactions between subnational regions and further pointing to the interlinkage between vertical and horizontal dimensions of policy interaction. A chain of drill-down effects triggered by the central government's policies leads to adjustments at the regional level. In most cases, this then either accelerates or hinders horizontal policy diffusion. Two varieties of top-down diffusion emerge, namely those driven by policies and measures related to the national ETS and those triggered by policies and arrangements beyond ETS (see Table 5.3).

Table 5.2 Bottom-up policy influence towards national ETS

Forms	*Driving party*
Hosting registry/trading platform and supporting development of national ETS rules	Regional government and exchange (national government with final say)
Co-development of national ETS rules	National government (in lead) and regional technical supporting institutions
Sending regional experts to national ETS preparation working group	National government
Commenting on draft national ETS policies	National government
Diaoyan	National technical supporting institutions
Sharing regional experience at centrally organized capacity building activities	National government
Engaging with national policymaking via international donor projects	Regional technical supporting institutions

Source: Author

Table 5.3 Top-down influence

Dimension	*Trigger*	*Drill-down effect*
National ETS policies	National ETS measures and technical guidelines	Alignment of ETS pilot policy design with national rules
	Prospects of national ETS gradually taking over key sectors within the ETS pilot	Future existence of ETS pilot in question; challenges regarding continued operation
	Assignment of national ETS capacity building center to regional pilot	Enhanced credibility of experience and knowledge transfer between pilot and other regions
	National decision to not expand ETS pilots but develop a centralized national ETS	Those originally pursuing regional ETS development via learning from pilots lose space/incentive
	Advancement of national ETS, including tasks assigned to regional governments	More non-pilot regions motived to learn from pilots' practical experiences
	Central government emphasis on avoiding overfinancialization of market products	Carbon market innovation constrained to regional level
	Government restructuring and slowdown of national ETS momentum	Slowdown of policy exchange and diffusion at regional level
Higher ranking/ broader policies and arrangements beyond ETS	Top-level policy documents and directional guidance from central government including new development concepts	Enhanced motivation of local governments to pursue ETS as a source of political achievement
	Other comprehensive/ policy-specific experimenting and piloting approved or promoted by the national government	Enhanced motivation of local governments to pursue ETS as a source of political achievement
	"Counterpart help" mechanism, i.e., pairing provinces	Pairing provinces extends policy exchange and cooperation also to climate, including ETS

Source: Author

Such far-reaching top-down influence on policy diffusion differs from Western countries, in particular those with federal systems where horizontal diffusion is more prominent. China's unitary system and the strength of its central government makes this unsurprising; however, the diversity of triggers and pathways of policy influence is unexpected and serves as a reminder of the complexity of China's broader governance as well as the multifaceted nature of its low-carbon transformation.

Indirect diffusion

An additional diffusion mechanism has been revealed by the empirical data of this study, unforeseen by the analytical framework developed based on existing mainstream diffusion theory: indirect diffusion. There are two sub-models of such ETS-to-non-ETS policy diffusion: the first, where policy A in the supplier region is followed by the development of policy B, which is then diffused into an external region; the second model, where the policy follower region uses policy A from the supplier region as a source of learning for their development of policy B. The triggers of policy change in the policy follower region functions similarly to the learning mechanism.

In the specific case of ETS in China as policy A, policies B include carbon and green finance, other market-based environmental or energy policies, and regional offset programs. The first model of indirect diffusion is more dominant in Shanghai, which pioneers carbon and green finance policy by also drawing on its ETS experience. This has led to the diffusion of this policy towards peer provinces or cities, such as Zhejiang and Dalian. The second indirect diffusion model is more relevant for Hubei, where other provinces, such as Jiangxi, have used Hubei's ETS experience to inform the design of their market-based ECQT. Indirect diffusion in Hubei is weaker than in Shanghai. More efforts are required for less-developed regions to drive policy diffusion towards other regions. Such investments must first be backed by strong political interest.

Indirect diffusion is enabled by factors similar to horizontal diffusion: geographical proximity, the compatibility of economic and industrial structures, and perception as a "leading region". Major forms of this diffusion are like that of learning, such as *Diaoyan*, workshops, and information channels. Like direct diffusion, indirect diffusion between regional governments is also largely influenced by national government policy and guidance, such as the promotion of green finance and ECQT policies. The overlap between technical supporting institutions (even when the government bodies in charge of the policies vary) makes these institutions instrumental in the intertwined diffusion processes across ETS and related policies. Finally, though the coexistence of multiple carbon- and energy-related trade policies has catalyzed indirect diffusion, this could potentially also lead to unnecessary policy competition that may hinder future policy diffusion. Mitigating this risk requires adequate national level policy coordination.

Time dynamics

Policy diffusion is a dynamic, complex, and intertangled process. The Shanghai and Hubei cases have revealed the time dynamics of ETS policy diffusion and the effects of changing exogenous and endogenous environments and regional players.

Competition emerged before the ETSs launched around 2013–2014, concerning which pilot would be the first to get their system up and running—or at least, which could avoid being the last (Interview, BJ-E-040220). Shanghai took second place. Indeed, between June 2013 (launch of Shenzhen's pilot) and the end of that year, four other pilots were launched (EDF & ERI, 2018).

For Shanghai, several diffusion mechanisms occurred in parallel in the design and early operation phase (up until around 2015). Most intensively, learning occurred between Shanghai and its pilot and non-pilot peers, the latter of which took more initiative in approaching Shanghai. For diffusion with other pilots, both sides were seen to be equally proactive. Post-launch, Shanghai made efforts to promote regional cooperation with its East China peers, demonstrating the carrot approach to the coercion mechanism.

In the design and early operation phase, Hubei focused on its own ETS development. "[At the beginning] we were a student . . . but we took advantage as a latecomer to draw on others' experiences and lessons" (Interview, HB-E-071219). Hubei focused on design, implementation, evaluation, and improvement until around 2015, after which increasingly more regions instead came to learn from their experiences, thanks to the pilot's sound market performance and high liquidity. Hubei went from being a "student", to passively responding to non-pilots' demand for policy advice, to more actively engaging with others to enhance its profile and influence.

Hubei also initially attempted to promote Central China regional ETS cooperation and explored linking with the Guangdong ETS pilot. It continued to encourage regional cooperation though with limited and short-lived success. In conjunction with extensive competition and learning, socialization was a minor diffusion mechanism in this phase, with a focus on promoting the success of the Hubei ETS.

Similar "peak times" emerged for the more dominant diffusion mechanisms in both Shanghai and Hubei. A spike in learning then occurred after the NDRC assigned the two regions to be national ETS capacity building centers in 2016. Hubei, for example, began to proactively approach non-pilot regions to help build government and private sector capacity. This in turn intensified competition between the pilots' regional carbon exchanges. Competition spiked again when the NDRC initiated the selection process for hosts of the national ETS market infrastructure.

Institutional and policy changes at the national level between the two FYP periods have impacted the pattern and pace of regional level policy diffusion. Firstly, the central government's 2018 reform package saw climate change responsibilities shift from the NDRC to the MEE, with a similar shift at the regional level to the local BEEs. This led to a decrease in overall political support for ETS policy innovation and diffusion in many subnational regions, a result of several factors: the realignment of working styles between the previous and new ETS authorities (e.g., that the DRCs focused more on macrolevel strategic thinking and innovation, while environmental agencies more on the micro- and operational level) and staff changes and lack of familiarity with the topic of carbon mitigation and market-based mechanisms; and priority and resource competition (e.g., between existing pollution control measures and new ETS responsibilities).

How this national ETS was to develop influenced both the type of participant involved in policy diffusion and the focus of the policy learning. In the mid-2010s, the central government announced the top-down and centralized development of the national system, as opposed to expanding the number of pilots. Before this, policy learning was driven by non-pilot regions interested in establishing their own ETSs; afterwards, many more regions got involved. Learning also shifted from focusing on the comprehensive design and operation of a functioning carbon market to the specific requirements set for regional governments in preparation for the national ETS, such as historical data collection.

The pace of the national ETS preparations impacted the speed of policy diffusion at the regional level. During the first half of the 13th FYP, the advancement of the national ETS catalyzed overall policy diffusion, providing impetus for diffusion between ETS pilots (like Shanghai and Hubei) and many non-pilot regions, as well as between pilots and the national government. Accordingly, the slowdown of national ETS momentum during the second half of the 13th FYP period hindered policy diffusion across subnational governments.

Institutions and actors

This book reveals how Chinese climate policy is institutionally organized and operated at the regional level. Identifying the principal governance structures and actors in policymaking and diffusion, as well as how they interact and are intertwined, allows for a better understanding of regional power structures. The regional institutional arrangements underlying China's climate and ETS governance are hugely diverse, mirroring the diversity of governance traditions and capacity (see Table 5.4).

In both Shanghai and Hubei, the regional government has worked closely with and relied extensively on the technical supporting institutions throughout

Table 5.4 Main institutional players in ETS governance: Shanghai and Hubei

Role	*Shanghai*	*Hubei*
Regulatory body	Shanghai municipal government, ETS competent authority, carbon exchange	Hubei provincial government, ETS competent authority, carbon exchange
Political guidance	Leading Group on Climate Change, Energy Conservation and Emissions Reduction	Leading Group on Energy Conservation and Emissions Reduction (Climate Change)[2]
Competent authority	Shanghai DRC (until 2018); Shanghai BEE (from 2019): Division of Atmospheric Environment and Climate Change	Hubei DRC (until 2018); Hubei Department of Ecology and Environment (from 2019): CCD
Centralized coordination	Leading Group on Shanghai ETS Pilot and its office	n/a
Technical supporting institution	SIC	Wuhan University's Climate Change and Energy Economics Study Center
	SEEE	CHEEX
	SECERC	CQC
Compliance enforcement agency	SECSC	Competent authority
ETS Expert Committee	Formed in August 2012; no public information available regarding composition	Formed in March 2016; composed of 13 experts from 11 regional and national institutions

Source: Author

all stages of their ETSs. Support is provided for regulation framework development and policy design of ETS elements, including cap-setting, allowance allocation, MRV, trading rules, compliance and market oversight, and review. These institutions have also helped to develop and manage market infrastructure, emissions reporting systems, and trading platforms, communicate and engage with business and other market participants, provide capacity building for sub-provincial authorities, verifiers, and business, and expand ETS to new sectors. Whereas Shanghai has relied on local technical supporting institutions, Hubei has drawn on expertise from a range of institutions both regional and national.

Institutional setups have evolved over time in both cases, with shifts in climate change responsibilities and subsequent changes in the ETS competent authority to adjustment of the leading groups. In Shanghai, the ETS leading group merged with that for broader climate change, energy conservation, and emissions reduction; Hubei's broader leading group instead swiveled its attention to climate change. The supporting institutions changed most notably. While the SCI gradually took on central coordination and daily operation in the Shanghai ETS, the SECERC has moved away from policy advisory to focus on providing verification services. Similarly, in Hubei, the CQC has gradually moved to acting as a third-party verifier, and Wuhan University has also shifted its focus from climate policy advice to academic research, gradually making the CHEEX the central supporting institution for the province. Institutions in both cases worked together more closely during the preparation and early operation phase and less so in the later phases. The more centralized arrangements such as the office in Shanghai and Hubei's thematic working groups were dismissed once the ETSs entered their stable operation phase.

The similar successes of Shanghai and Hubei in diffusing ETS policy, despite their divergent geographical locations and governance traditions, can be attributed to important similarities between their institutional structures. These include strong political support and commitment from regional leadership, high institutional and individual stability, dedicated financial resources, and a considerable level of transparency. Both have secured high-level leadership support above the immediately responsible government body, personnel from the competent authority and supporting institutions kept highly stable, and both with special funds to channel resources from the regional budget to support ETS.

The empirical data has made clear that the government (the ETS competent authority) and its main technical supporting institutions are the two dominant types of institutional players. Table 5.5 indicates the level of involvement of these players across the diffusion processes.

Table 5.5 Institutional players' involvement in ETS policy diffusion: Shanghai and Hubei

Diffusion type	*Shanghai*				*Hubei*			
	Gov.	*SIC*	*SEEE*	*SECERC*	*Gov.*	*CCEE*	*CHEEX*	*CQC*
H: learning	xxx	xx	xxx	x	xxx	xx	xxx	x
H: competition	xxx		xx		xxx		xx	
H: socialization	xx	xx	xxx	x	xx	x	xxx	x
H: coercion	xxx		xx		xxx		xx	
V: bottom-up	xxx	xx	xxx	x	xxx	x	xxx	x
V: top-down	xxx		xx		xxx		xx	
I	xx		xxx		xxx		xx	

Source: Author

Note: H: horizontal diffusion; V: vertical diffusion; I: indirect diffusion. Xs indicate the level of involvement: 3 strong, 2 medium, 1 minor, and 0 imperceptible/no available data.

Technical supporting institutions' expertise and close *Guanxi* with decisionmakers contribute to their important role in diffusion. Previous involvement in market-based mechanisms, such as the CDM and energy conservation policies, laid the foundations for extensive ETS knowledge. Newly established carbon and environment exchanges or trading centers have unique expertise in market operation. *Guanxi* can be traced by looking at the institutional types and founding histories. As has been well documented in literature on Chinese governance, the significance of *Guanxi* is a unique feature of its domestic policy diffusion, as compared to the West.

These institutions' involvement in policymaking and diffusion has evolved as ETS developments unfold. While several share the balanced role of supporting the regional government at earlier ETS phases in both case studies, the SIC (a government-affiliated research institution) in Shanghai and the CHEEX (the more market- and profit-driven carbon exchange) in Hubei have become the most "dominant" supporting institutions for policymaking and administration in their respective regions.

Beyond these two regions, several institutions in some instances share the supporting roles across different phases in a balanced manner, such as the Beijing Climate Change Research Center (government-associated), the China Beijing Environment Exchange (carbon exchange), and Tsinghua University (academia). In other cases, one supporting institution is more dominant. However, dominance in jurisdiction internal ETS policymaking does not necessarily equate to the same for policy diffusion to other jurisdictions. Instead, across both empirical cases and in other regions, carbon and environment exchanges have been particularly active in ETS policy diffusion (see Table 5.5), driven by business interests. Institutions significantly impact policy diffusion when they support regional governments with climate, energy, or environmental policies.

Institutional players at the national level have also been seen to be instrumental. The central government's support for institutions such as the NCSC (a government-associated institution, previously affiliated with NDRC and since 2018 the MEE), Tsinghua University, SinoCarbon, the CEC (a technical power sector association with SOE membership), and the CQC (a quasigovernment-associated institution) has been pivotal in vertical policy diffusion processes. Most play a dual role of providing services and support to both the government and the market. Several, such as SinoCarbon and the CQC, have also directly supported subnational ETS development, thus also facilitating horizontal policy diffusion.

These technical supporting institutions show a remarkable degree of path dependency and institutional resilience. Once institutions are established or become involved in the ETS policy process, they become advocates of the instrument's development and diffusion, fueled by their own institutional interests. The level of this advocacy varies; institutions taking the form of consultancies and exchanges are more active than nonprofit bodies such as government-affiliated institutions and academia.

Companies covered by the ETS and sector associations are also important. Among them, SOEs are a pillar to the Chinese economy, usually occupying strategic sectors as large entrepreneurs. They hold a unique position in the governance structure; most SOE executives have an administrative rank, granting them greater access to inner government circles. Shanghai and Hubei differ in terms of government-business relationships. In Hubei, large SOEs and industrial associations act in a similar way as lobbyists in Western countries, while in Shanghai the government is more powerful with well-organized channels and rule-based processes to engage with industry. The strong bargaining and lobbying behaviors of Hubei's private sector stem from its sheer market power, the administrative level-generated central SOE problem, local protectionism, and a less modern governance tradition.

Despite these differences, how the government engages and communicates both formally and informally with the private sector is similar: *Diaoyan*, requests for comment, discussion symposiums, and training. However, consultation processes in Shanghai are clearer, more transparent, and regular, which leaves less space for rent seeking and exchange of favors as compared to Hubei.

SOEs are heavily involved in ETS agenda setting. However, the empirical data has shown that their role in the further diffusion of ETS outside of their operational scope whether horizontally or vertically is limited and passive. SOE priorities lie in policy engagement within regional boundaries and ETS coverage, as well as internal strategies to respond and comply with the carbon market. In some cases, private sector representatives are invited by the regional government or leading supporting institution to share their experiences with other regions as part of the learning diffusion process.

Although civil society actors like the media, NGOs, and the public are considered to be important stakeholders in policymaking and diffusion in the Western context, their role in China's ETS governance and policy diffusion is limited. This is in part due to the concept of "Big Government, Small Society".

Chinese media

The media landscape and its role in climate governance in China is particularly complex. Most traditional media outlets are government-owned. There are three main types of media relevant for climate change coverage: official, state-run media who set the agenda for other media outlets; newer, more independent media who are freer in their reporting and operate in a more competitive market; and online portals. There are no specialized climate media outlets, though these exist for energy.

Chinese coverage of climate change is often marginalized and concentrates on UN negotiations. Its focus on topics such as low-carbon policy, technology, and business is greater but still fragmented. This is because climate is only a small part of the main reporting disciplines, which include political and industrial economy. Other more "tangible" environmental issues such as air pollution appear more prominently in public debate. Secondly, journalists tend to report on climate change as a minor and temporary topic and thus lack depth of knowledge. Thirdly, climate change is often perceived not to impact everyday life; news thus targets only niche or elite audiences.

Broader political circumstances also weaken the media with regard to climate change. There is a constant struggle between the need for informational freedom and the government's desire to control content via censorship. The media also lacks the professionalization essential to shaping public debate (Interview, N-M-100220). The Chinese government has tried to use the media to sway international and domestic public opinion on climate.

Though China's media landscape is slowly undergoing digitalization, no substantial change in its limited influence is expected in the near future. Despite all this, there is still space for individual media and journalists to influence domestic climate policymaking via quality reporting and partnerships with other institutions.

Other players

Unpacking how such diverse players interact reveals the power structures that lie behind regional ETS governance. The government is the most powerful and central. Technical supporting institutions come second place. With direct access to and strong connections with the government, they are the bridge between business and government. Businesses, on the contrary, are weaker and have more limited access, though within this group SOEs have easier and more direct access. Regional differences also emerge; business is more powerful in Hubei than in Shanghai. Civil society has the least gravity.

Individual policy entrepreneurs in China, like in Western democratic systems, are also widespread and pivotal, though they are mainly based inside of the government. They exist on a continuum from career bureaucrats at the departmental level all the way through to provincial leadership. These individuals share certain qualities, including the ability to learn quickly, high professionalism, and strong managerial skills. Several are also particularly politically courageous. Their expertise and the value they attach to ETS have allowed them to advance policy development, successfully advocate for it, and deliver policy innovations.

Conceptual implications

This book contributes to three pillars of research: firstly, the application of the conceptual framework of policy diffusion to China; secondly, the ongoing debate on China's governance; and thirdly, the relatively young transformation governance literature on China.

Applying a comprehensive policy diffusion framework to China

Since its origin in the 1960s with a focus on state-to-state diffusion within the United States, policy diffusion theory has continued to evolve on both theoretical and empirical fronts. However, most existing empirical studies thus far only test one diffusion mechanism. Existing literature has also focused on Western systems. Empirical research on China and other emerging economies is still scant. Applications of diffusion theory to China's policy processes focus on the impact of international factors. This book instead applies the analytical framework to China's domestic governance and the influence of Chinese actors on innovative policy deployment and explores a range of diffusion mechanisms.

We further expand the explanatory power of policy diffusion theory from Western democratic systems to the Chinese governance system. Its applicability to authoritarian regimes is subject to careful testing, and the research presented here has confirmed the findings of other recent policy diffusion studies.

Testing the policy diffusion framework by applying it to China also reveals some of its limitations. For example, it is difficult to draw distinct lines around the microlevel characteristics between the diffusion mechanisms, as there are other factors that change the pattern of these characteristics. It is also difficult to collect sufficient empirical data to make such a comparison, due to the relativity of the differentiation (a general challenge) and data availability (a challenge more specific to China). To counter this, this book moved beyond purely defining specific diffusion mechanisms at play by investigating micro-level characteristics and applying detailed process tracing backed by extensive expert interviews.

The (de)centralization debate

Scholarly debate is ongoing regarding China's governance and its suitability for the modernization and transformation of the economy. This book shows that both centralized and decentralized structures exist in parallel. In many instances, they are mutually reinforcing. Indeed, this research shows that centralization and decentralization are so intertwined that it is unreasonable to peer only at one. It is also not helpful to portray the two contradicting each other and neglect their positive interactions, particularly in the context of climate governance.

Another debate is on the *quality* of China's governance. As Francis Fukuyama (2013) posits, the existing measurements of China's quality of governance are far from adequate, which often leads to a biased low rating. Such underestimation of China's quality of governance is partially based on value judgments and technical problems. The merits of focusing on capacity (i.e., resources and the degree of staff professionalism) in discussing quality of governance are obvious. We must also disaggregate states "into their component parts, both by function, region, and level of government" (ibid., p. 16).

Scholars have historically tended to view decentralized governance as a solution to the modern state. This book demonstrates the opportunities provided by governance models that may not necessarily fit the typical decentralization ideal. Progressive Chinese regions with forward-looking agendas that pioneer policies and influence their peers to follow a similar trajectory, the institutional players contributing to or participating in the policymaking and implementation process, the policy entrepreneurs, and the informal channels between governments are important tiles of China's governance mosaic. China is vast and replete with layers in its government and many more in its society. Robust framework conditions are required for a fact-based research agenda, for which access to information and transparency are essential.

Transformation governance with Chinese characteristics?

The transformation from the current development pathway towards a carbon-neutral one is a complex and demanding task for all societies. Good governance is central to transformation; however, emerging research on the matter has spiraled around the "what" and "why", with empirical analyses restricted to the national level and a few coastal provinces. This book unpacks two regions' experience of the "how" and "why" of ETS and ETS policy diffusion.

How much is China different from the West? We are left with a mixed picture. Despite stronger vertical dimension in addition to horizontal (unlike the West), in many cases, top-down influence injected by the Chinese central government provides positive impulses for policies to transfer horizontally from one region to another.

Two more similarities highlight the potential of a coherent research approach to studying transformation governance across developed and emerging economies. Firstly, the empirical cases have illustrated the importance of policy entrepreneurs in China's domestic policymaking and diffusion. Secondly, like the "laboratories of democracy"[3] that exist in Western countries, many subnational regions in China are "laboratories of good governance". Not only are the eastern coastal regions leaders in various economic, social, and environmental policies under more modern and open governance models, but regions in the rest of China are catching up, many explicitly benchmarking themselves against their eastern peers to accelerate this process. These laboratories of good governance may be further cultivated to become "laboratories of transformation".

Practical implications

Policy diffusion also has important practical implications (Gilardi, 2016). The evidence shows that when policy practitioners in China maintain space at the regional level to experiment, innovative new policies are more likely to diffuse across regional borders. This space has been preserved in the past largely via piloting mechanisms but would likely benefit from being mainstreamed as standard practice. A political culture that encourages trying new measures without penalization if things go wrong must be coupled with the right incentive mechanisms. In this regard, the central government's recent tightening of control over the subnational governments may have potential negative impacts on regional innovation, learning, and virtuous competition.

The MEE is traditionally a weak ministry with a focus on end-of-loop policies and command-and-control. Now, it is faced with the opportunity to embrace the more strategic and innovation-oriented approach of the NDRC. A greater focus on market-based policies that not only enhance local governance capacity but also hold potential for policy learning is also necessary. However, the interviews underlying this book reveal that such MEE changes may not come about easily without external steering, for example from the central government.

Guidance from the central government is an important catalyst (or necessary condition) also for policy innovation and diffusion at the regional level. The concepts of ecological civilization, high quality development, and green development are likely to yield better results if formalized via policies, performance evaluation mechanisms, and institutions. The climate and energy performance evaluation mechanism already in place could also have room for improvement regarding its accountability framework and if it were to swivel its focus towards more transformative policies like ETS.

China may benefit from other innovative institutions with a high-level mandate that could set and monitor national aggregate or sector-/region-specific climate targets. International examples, such as the climate change cabinet in Germany, the committee of climate change in the UK, and the climate change commission in New Zealand, may prove useful inspiration.

An importance difference lies between attention given to the "message" (e.g., real stories highlighting the economic and social co-benefits of "green" policies) and the "messenger" (e.g., regional governments from leading regions). For example, both the USA and Europe have been important drivers of China's climate transformation as Chinese (especially government) perception on climate change is hugely influenced by their decisions and messaging. A tailor-made approach has the potential to help "internalize" the concept across levels and regions in China, something that currently is led only by a small group of policy elites.

China's experience is also instrumental in accelerating the global transformation, especially as an example for other developing and least developed countries. The country has already begun to invest in efforts to share its policy experiences. However, it currently employs both bilateral and multilateral approaches in its international outreach. Prioritizing the multilateral approach would differentiate China from the USA and align well with China's "responsible great power" strategy. Greater transparency in its bilateral cooperation would enhance China's credibility and international image.

There is also much space for the Chinese technical supporting institutions to contribute to international cooperation. For the Chinese approach to move beyond using its economic weight to influence international partners and thus secure more resilient partnerships, non-governmental players may also complement the dominance of government voices.

Climate and economic transformation set among the top priorities for international cooperation with China would likely alleviate some of the volatility of domestic and international politics. The evidence shows that working with China at both national and subnational levels leads to the most fruitful outcomes. Not only the eastern coastal areas are pertinent; regional "leaders" elsewhere, such as Hubei for Central China and Chongqing and Sichuan for Southwest China, are also significant. Support towards these regions would accelerate their transformation and facilitate progressive regional climate partnerships.

It is also clear that policy entrepreneurs or those with such potential can greatly influence the trajectory of climate policymaking. Some international players may be in a better position to work with these individuals on a project basis, helping to accelerate their policy agendas. Others may help connect them in meaningful networks. Increased international exposure via fellowships or exchange programs can also be pivotal. Well-aligned external strategies for communicating and coordinating with Chinese policy entrepreneurs could well lead to desired outcomes.

The special position of Chinese SOEs within the economic and political landscape is unlikely to change in the foreseeable future and thus requires at the institutional level carefully designed engagement strategies, especially with the more progressive of these.

Finally, the development and expansion of international networks and initiatives—for instance, trilateral cooperation involving one developed country, China, and one developing country—may also yield success.

Notes

1 The year 2018 slowed momentum for diffusion, but this is left out of the graphic for simplification.
2 "Leading Group on Climate Change, Energy Conservation and Emissions Reduction" since 2019, indicating a more balanced focus on climate change.
3 A concept that explains how within the federal framework, there exists a system of state autonomy where state and local governments act as social "laboratories", where laws and policies are created and tested at the state level in a manner similar (in theory, at least) to the scientific method.

References

EDF & Energy Research Institute (ERI). (2018, May 23). *The Progress of China's Carbon Market 2017*. www.edf.org/climate/report-evaluates-chinas-national-carbon-market

Fukuyama, F. (2013). *What Is Governance?* (Center for Global Development Working Paper No. 314). https://ssrn.com/abstract=2226592

Gilardi, F. (2016). Four ways we can improve policy diffusion research. *State Politics & Policy Quarterly*, 16(1), 8–21.

6 Recent developments and a view ahead

Policy diffusion is an interactive process. This book's detailed empirical case studies reveal the coexistence of both horizontal and vertical diffusion in the field of transformative climate policy in China. Since the 2011–2019 period covered by these case studies, ETS in China has advanced, with the launch of the Chinese national ETS in early 2021. What has changed regarding ETS policy diffusion in this new post-piloting phase? What next?

The influence of national policies and politics

In this new period, the powerful influence of national level policy and politics—top-down vertical diffusion—is clear. Firstly, the Chinese national ETS has significantly impacted the sectoral coverage of the regional ETS pilots. The system regulates more than 2,000 companies from the power sector (including CHP and captive power plants in other sectors), which emit more than 26,000 tCO_2 per year (ICAP, 2022a). According to the trial Carbon Emissions Trading Management Measures released in January 2021, compliance entities that are covered in the national ETS will no longer participate in the local ETSs. The National Carbon Emissions Trading Market Construction Plan for the power generation sector, released in December 2017, stipulated that the pilot ETSs would gradually transition to the national ETS, while simultaneously continuing to explore design and development options for the national ETS (NDRC, 2017). In practice, over the course of 2021 and 2022, all regional pilots gradually transferred their power sector entities to the national ETS, retaining other sectors in their regional markets. To ensure adequate market size, many pilots have also expanded to other sectors (ICAP, 2022b).

Secondly, the launch of the Chinese national ETS put an implicit stop to the development of any new regional systems. Shenyang city in Liaoning province in northeast China has made an attempt—in late August 2021, the Shenyang municipal government issued Measures for the Management of Carbon Emissions Trading in Shenyang, which provide the general policy framework and legal basis for a regional ETS, to take effect the following month (ICAP, 2021). However, due to lack of national political support, preparations seem

DOI: 10.4324/9781003325307-6

to have come to a halt; no information on the further development of the Shenyang ETS has been made publicly available since.

The Chinese national ETS has positively influenced the innovation and use of carbon offsets in subnational regions. With the genesis of the national carbon market, the temporarily suspended CCER mechanism—and its relaunch—has returned to the spotlight (Qiu, 2022). This has led to innovation and experimentation from local pilots regarding offsetting emissions. For example, Guangdong, Shenzhen, and Chengdu have launched their own Tan Pu Hui offset products; Beijing has begun trading green travel emissions reductions, and Fujian has introduced carbon sinks for marine fisheries and tea plantations (ibid.).

Finally, shifts in the climate governance structure at the national level have also impacted regional ETS policy diffusion. After setting the national "dual carbon" targets (peaking emissions and achieving carbon neutrality) in 2020, the Chinese government shifted its climate governance structure once again. The NDRC is back in charge of target-setting and coordination for the dual carbon policies, while ETS remains under the remit of the MEE. As a superministry, the NDRC has added fuel to China's climate policy, releasing in 2021 and 2022 a series of documents to underlie the existing dual carbon policy framework. However, ETS is one of the few climate issues still managed by the less powerful MEE. This has slowed momentum for the development of the national system and ETS policy diffusion to non-pilot regions.

Progressive regions continue to expand their carbon markets and interact with others

Despite the significance of vertical diffusion, China continues to also display horizontal policy diffusion among pilot regions. The launch of the Chinese national ETS did not mean the end of the regional carbon markets. All are still in operation, and many have even expanded their scope and continue to innovate and actively engage with other regions.

For example, Guangdong is working to advance its ETS in collaboration with others, in particular its neighboring regions. According to the Guangdong Province Work Plan for the Construction of Ecological Civilization During the 14th FYP released in November 2021, the Guangdong ETS plans to expand its compliance market's sectoral scope, as well as explore the development of its regional offset program. In addition, Guangdong will also research the feasibility of the construction of a joint or linked carbon market for the Guangdong-Hong Kong-Macao Greater Bay Area (ICAP, 2022c). Tianjin is also working to improve its ETS, for instance by strengthening compliance mechanisms. It is also introducing allowance auctioning (ICAP, 2022d). As a market based initially on free allocation, the Tianjin ETS held its first auction in 2020 and two auctions each in 2020 and 2021 (ibid.). In September 2021, the Standing Committee of the Tianjin Municipal People's Congress issued the Tianjin

Carbon Peaking and Neutrality Promotion Regulations, which for the first time formally introduced financial penalties for noncompliance (ibid.).

Two main factors contribute to this continuation of horizontal diffusion. First is the regional governments' interest in preserving the resources they spent on their systems and the institutions such as the regional exchanges that have been successful and beneficial (Interview, HB-G-250220). Both Hubei and Shanghai, for example, established new institutions, either independently or as part of existing institutions. These institutions often have close relationships with the regional governments and became "ETS champions", both within their regions and beyond. The relative success of ETS as a mitigation tool against a backdrop of tightening regional climate targets over the coming decades also drives strong government interest in the instrument. As such, regional governments want to continue benefiting from ETS, with as much control as possible in their own hands (Interview, TJ-E-040220).

The second reason is that ETS has natural linkages with other policies such as green finance, carbon finance, and carbon offsetting. High momentum for these other policies, often prompted by more powerful ministers such as the Ministry of Finance, also provides motivation for regional governments to leverage the potential of ETS in these areas. ETS acts as a backbone for green and carbon finance innovation. For example, the Shanghai municipal government has expressed the city's commitment to "make good use of the advantages of operating the national ETS trading platform . . . and actively promoting carbon finance innovation . . . Shanghai [should be built] into a carbon trading, pricing, and innovation center with international influence" (People's Daily, 2022). Compliance carbon markets also provide demand for carbon offset credits. In regions with great interest in promoting offsetting, it makes sense that they are also keen to strengthen their ETSs.

External influences on new dynamics in China's ETS policy diffusion

Unlike the piloting phase, this post-pilot period sees external factors—policies outside of China—injecting new dynamics into China's domestic carbon pricing policy diffusion. The most prominent of these is the carbon border adjustment mechanism (CBAM) proposed by the EU.[1] In December 2020, the EU announced ramped up climate targets: reducing the bloc's GHG emissions by at least 55% by 2030 below 1990 levels and achieving carbon neutrality by 2050. This was a substantial increase in ambition compared to the previous target of a 40% reduction by 2030. To deliver on these promises, the EU's climate policy framework required bolstering—in particular the backbone of the framework, the EU ETS. Key elements included tightening the cap and phasing out the number of allowances freely allocated to emitters covered by the EU ETS.

This in turn raised the question of potential carbon leakage. Carbon leakage is where, as a result of stringent climate policies such as carbon pricing, companies move their production abroad to jurisdictions with less ambitious climate measures in place. This can lead to a rise in global aggregate GHG emissions, in effect cancelling out the effects of the stringent policies in the first place (PMR & ICAP, 2021). To mitigate this risk and protect industry competitiveness, the EU then proposed alongside the revision of the EU ETS the implementation of a CBAM, which it laid out in its European Green Deal legislation package released in July 2021 (European Commission, 2021). The EU CBAM is set to put a price on carbon emissions embedded in goods imported into the bloc, ensuring that industries in other countries exporting to the EU are also faced with a carbon price that is comparable to the EU Allowance price. In December 2022, negotiators of the Council of the EU and the European Parliament concluded a trialogue process and reached a provisional agreement on the CBAM.

Though it took the EU over a year to conclude the legislative process, since being proposed the CBAM has catalyzed carbon pricing developments in several countries outside of the EU, including in China. As an external factor on trade, it motivates domestic policy discussions in other countries on how to establish new or strengthen existing carbon pricing instruments, including ETS. For example, the EU CBAM has piqued the interest of policymakers and stakeholders in countries in the Asia-Pacific region, such as Indonesia and Thailand, to develop their own carbon markets as a preemptive countermeasure. In China, it has revived discussions around improving its own national carbon market and potentially linking the EU's and China's ETSs in the mid to long term (Kardish et al., 2021).

The CBAM has also accelerated talk of a carbon tax as well as hybrid carbon pricing instruments. Among other Chinese climate and energy experts, Shao and Xu (2022), for example, note that the level of carbon prices in China is still low, even to a degree of market failure. This makes it difficult to effectively incentivize companies to reduce emissions and participate in carbon trading activities, and thus also difficult to achieve the desired effect of reducing emissions on aggregate. Introducing a carbon tax—where a price level (rather than volume, as in an ETS) is set—can be one way to effectively guide emissions reductions in sectors not covered by the ETS and to alleviate the problem of low carbon prices. The topic of a potential carbon tax has also surfaced in discussions at the Bo'ao Forum for Asia, a high-level international conference backed by Chinese political leadership that is held annually in Bo'ao, Hainan Province (Bank of China Insurance News, 2022). Furthermore, in 2022 the NRDC launched tenders for 13 research projects that would work on policymaking to support peaking carbon emissions and reaching carbon neutrality. A consortium led by the Chinese Academy of Fiscal Sciences (CAFS) won one of these projects that focuses on carbon pricing (NDRC, 2022). The CAFS is a national think tank associated with the

Ministry of Finance, which oversees taxation. That such an institution won hints at the possible revival of a carbon tax in China.

The dawn of a new era

Beyond ETS policymaking and governance, China is undergoing new reforms directly influenced by top leadership: President Xi Jinping. Significant changes came out of the weeklong 20th National Congress of the CCP that took place in late October 2022, which will impact Chinese policy and politics in the years to come. Xi has secured a historic third term ruling China, and the CCP has constitutionally entrenched Xi as "the core leader of the Central Committee". The new top leadership is comprised entirely of Xi's allies, and the Party Congress's work report is littered with ideas, such as the "new great struggle" that will come about as the country moves towards prosperity, and issues of national security (ASPI, 2022; Rudd, 2022). On the green agenda, mention of China's decarbonization goal at the 20th National Congress shows it has high-level buy-in (Xue, 2022).

In this way, China is exhibiting seemingly contradictory trends: tightening political control and securitization lie alongside commitments to green development. The profound effects of this juxtaposition on policy diffusion will continue to gain momentum and are likely to impact the rate of diffusion at the subnational level. These impacts will be important to follow closely, as the world—and China with it—enters a new era.

As the planetary pendulum swings towards increasingly intolerable climates, governments on various levels must leverage each other's experiences and exploit the diffusion of policy so that it can help accelerate climate action. China's experiences have shown in what settings this can work. The pace and scale at which this must take place will dictate how the future unfolds—and how livable it will be for the generations to come after us.

Note

1 Besides the EU, other countries are also developing their own CBAMs. For example, in June 2022, four US senators jointly introduced the Clean Competition Act, legislation (not yet concluded) aimed at making American companies more competitive in the global marketplace and tackling GHG emissions.

References

Asia Society Policy Institute. (2022). *Decoding the 20th Party Congress*. https://asiasociety.org/policy-institute/decoding-chinas-20th-party-congress

Bank of China Insurance News. (2022). 碳税与碳交易结合 "定价" [Combining Carbon Tax and Carbon Trading with "Pricing"]. www.cbimc.cn/content/2022-04/24/content_460312.html

European Commission. (2021). Proposal for a "Regulation of the European Parliament and of the Council, Establishing a Carbon Border Adjustment Mechanism". Brussels, 14.7.2021. COM (2021). 564 final. 2021/0214(COD). https://eur-lex.europa.eu/legal-content/EN/TXT/?uri=CELEX:52021PC0564

International Carbon Action Partnership (ICAP). (2021). *China's Shenyang Prepares to Launch a Local ETS*. https://icapcarbonaction.com/en/news/chinas-shenyang-prepares-launch-local-ets

ICAP. (2022a). *China National ETS*. https://icapcarbonaction.com/en/ets/china-national-ets

ICAP. (2022b). *Emissions Trading Worldwide: Status Report 2022*. Berlin. https://icapcarbonaction.com/en/publications/emissions-trading-worldwide-2022-icap-status-report

ICAP. (2022c). *Guangdong Pilot ETS*. https://icapcarbonaction.com/en/ets/china-guangdong-pilot-ets

ICAP. (2022d). *Tianjin Pilot ETS*. https://icapcarbonaction.com/en/ets/china-tianjin-pilot-ets

Kardish, C., Duan, M., Tao, Y., Li, L., & Hellmich, M. (2021). *The EU Carbon Border Adjustment Mechanism and China: Unpacking Options on Policy Design, Potential Responses, and Possible Impacts*. Berlin: Adelphi. www.adelphi.de/en/publication/eu-carbon-border-adjustment-mechanism-cbam-and-china

National Development and Reform Commission (NDRC). (2017). 国家发展改革委关于印发《全国碳排放权交易市场建设方案（发电行业）》的通知 [The NDRC Issued the "National Carbon Emissions Trading Market Construction Plan (Power Generation Industry)" Notice]. www.ndrc.gov.cn/xxgk/zcfb/ghxwj/201712/t20171220_960930.html?code=&state=123

NDRC. (2022, August 17). 关于确定2022年度碳达峰碳中和课题项目委托研究承研单位的公告 [Announcement on Determining the Commissioned Research Unit for the 2022 Carbon Peak Carbon Neutralization Project]. www.ndrc.gov.cn/xwdt/tzgg/202208/t20220817_1333102.html?code=&state=123

Partnership for Market Readiness & ICAP. (2021). *Emissions Trading in Practice: A Handbook on Design and Implementation*. Second Edition. Berlin. https://icapcarbonaction.com/en/publications/emissions-trading-practice-handbook-design-and-implementation-2nd-edition

People's Daily. (2022, July 27). 上海碳市场已纳入300多家企业，总体交易规模领先，这项纪录全国唯一 [Shanghai Carbon Market Includes More Than 300 Companies, Leading the Overall Transaction Scale, Only Record in the Country]. http://sh.people.com.cn/n2/2022/0727/c138654-40055747.html

Qiu, X. (2022). 2022碳中和行业最大悬念揭晓：CCER市场将重启 | 焦点分析 [The Biggest Suspense in the Carbon Neutral Industry in 2022 Is Revealed: The CCER Market Will Restart | Focus Analysis]. 36Kr Holdings Inc. https://36kr.com/p/1787488956854664

Rudd, K. (2022). The return of red China: Xi Jinping brings back Marxism. *Foreign Affairs*. www.foreignaffairs.com/china/return-red-china?fbclid=IwAR37XPEP8l64uUwd-iEbCqKd03P1wxMDw0X4eHCUMSU2b2f_VdlxilzUc4U

Shao, S., & Xu, L. (2022). 构建碳税与碳交易协同互补机制 [Building a synergistic and complementary mechanism for carbon taxation and carbon trading]. *China Social Science Network—Chinese Journal of Social Sciences*. http://chinawto.mofcom.gov.cn/article/br/bs/202205/20220503312833.shtml

Xue, Y. (2022, October 21). Climate change: China's Xi Jinping affirms net-zero commitment while touting coal's near-term value for energy security. *South China Morning Post*. www.scmp.com/business/article/3196741/climate-change-chinas-xi-jinping-affirms-net-zero-commitment-while-touting-coals-near-term-value

Appendix

Example interview

1) Briefly introduce your institution and yourself, including your role in your region's ETS policymaking process.
2) Describe your region's overall ETS policy development. What are the main objectives from the government perspective? What are the major achievements and challenges?
3) Describe the institutional setup and important actors involved in ETS governance in your region, including any changes due to the institutional reform. E.g.,

    a) Any specific institution(s) created?
    b) What are the main technical supporting institutions?
    c) Which companies or industry organizations have played an important role? How have government-business interactions occurred?
    d) Who are the major individuals pushing policy?
    e) What role have the media, NGOs, and the public played?

Policy diffusion mechanisms

Your region

Answer the following questions for the two phases below (alternatively, choose the phase you are more familiar with).

- *Phase One: start of the pilot until 2015*
- *Phase Two: 2015–2019*

1) Which other regions have Hubei/Shanghai influenced with regards to ETS policy adoption or implementation?
2) Who has taken the initiative here and what are their motivations?

3) Which of these measures best inject policy influence? If you list multiple measures, specify which are most important.
    a) Sharing the success and failure/cost and benefits of the ETS in Hubei/Shanghai been shared (If so, via which channels?)
    b) Promoting the concept of ETS to key policymakers in the region? (If so, via which channels?)
    c) Pressure from other region(s) regarding meeting energy or emissions reduction targets
    d) Providing specific incentives for other regions to develop their own ETS or promoting regional cooperation
    e) Hubei/Shanghai being viewed as "role model" for other regions (What is the impact of such perceptions?)
    f) Other regions experiencing "peer pressure" resulting in them learning from Hubei/Shanghai
4) What are the main outcomes of these measures? What policy impacts have been achieved regarding agenda-setting, policy design, or implementation?
5) What are the key institutions (individual institutions and networks) and people in this policy interaction process? What specific role has your institution played in this regard?
6) What are the success factors for and major challenges to injecting policy influence?
7) What policy impacts have Hubei/Shanghai injected into national ETS development? What are the main motivations? Compared to influencing policies at the peer province or city level, which are more important for Hubei/Shanghai?
8) In addition to ETS-to-ETS policy diffusion, describe any spillover effect between ETS and other relevant policies.

Comparison

1) Are there any changes across two different phases in motivations, measures, or actors of the policy interaction between Hubei/Shanghai and other regions? Why?
2) Are there any new regions that Hubei/Shanghai has injected policy influence in the second phase? If so, why here and which measures have been taken?
3) Across these two different phases, are there any typical cases of policy interaction and diffusion you would recommend as further case study? Why?

Beyond your region

Illustrate answers to the following questions with examples where possible.

1) Which pilots' regional governments are more active or successful in promoting exchange and learning about ETS (both between pilots and between pilots and non-pilots)?

2) In addition to local governments, what other institutions (such as exchanges, scientific research institutions, and consulting institutions) have performed well in the policy spill-over processes?
3) Which regions have played a positive role in the preparation and construction of the national carbon market?
4) How do national carbon market players (including government and technical support units) absorb local experience and lessons and via which channels?

Index